Dimensions Math
Tests 1B

Author
Dawn Yuen

Singapore Math Inc.

Published by Singapore Math Inc.

19535 SW 129th Avenue
Tualatin, OR 97062
www.singaporemath.com

Dimensions Math® Tests 1B
ISBN 978-1-947226-48-7

First published 2019
Reprinted 2020 (twice), 2021 (twice), 2022, 2023

Printed in China

Acknowledgments

Design and illustration by Cameron Wray with Carli Bartlett.

Preface

Dimensions Math® Tests is a series of assessments to help teachers systematically evaluate student progress. The tests align with the content of Dimensions Math K–5 textbooks.

Dimensions Math Tests K uses pictorially engaging questions to test student ability to grasp key concepts through various methods including circling, matching, coloring, drawing, and writing numbers.

Dimensions Math Tests 1–5 have differentiated assessments. Tests consist of multiple-choice questions that assess comprehension of key concepts, and free response questions for students to demonstrate their problem-solving skills.

Test A focuses on key concepts and fundamental problem-solving skills.

Test B focuses on the application of analytical skills, thinking skills, and heuristics.

Contents

Chapter	Test	Page

BLANK

Name: _____

Date: _____

25 min **Score**

| 30 |

Chapter 10 Length

Section A (2 points each)
Circle the correct option: **A**, **B**, **C**, or **D**.

1 Which candle is the tallest?

J   K   L   M

A M   **B** L

C K   **D** J

2 Which tape is the shortest?

W
X
Y
Z

A Z   **B** W

C X   **D** Y

3 Which side is the longest?

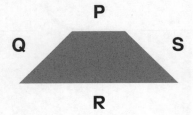

A P

B R

C Q

D S

4 Which ribbon is about 6 units long?

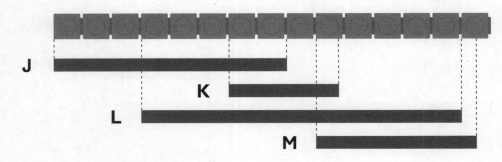

A J

B M

C K

D L

5 The chopstick is _____ 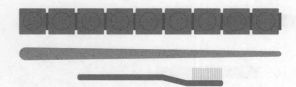 longer than the toothbrush.

A 1

B 5

C 6

D 4

Section B (2 points each)

6 Which side of the triangle is the shortest?

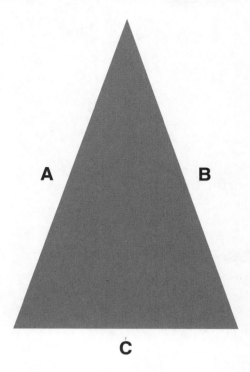

Side _____ is the shortest side.

7 Cross out the cabinet that is taller than Cabinet A and shorter than Cabinet D.

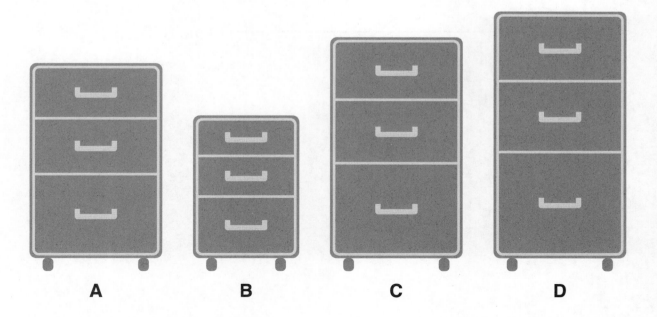

8 Cross out the longer string of beads.

9 Tape _____ is longer than Tape A and shorter than Tape C.

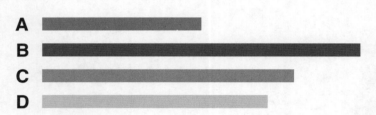

10 Draw a line that is shorter than Line A.

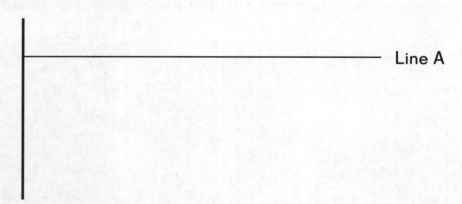

Line A

11 Which two trees have the same height?

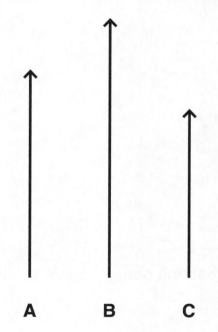

1 unit

A

B

C

D

Tree _____ and Tree _____ have the same height.

12 Put the arrows in order, from shortest to longest.

A   B   C

_____ _____ _____
Shortest

Look at the picture. Answer questions 13 to 15.

Use 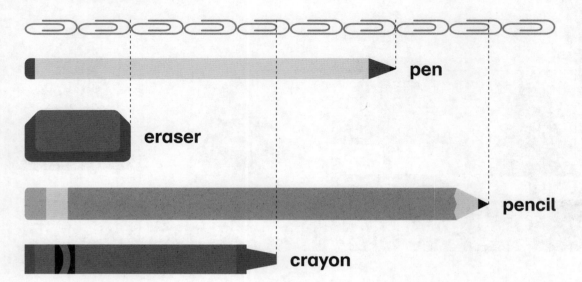 as 1 unit.

13 The _____ is shorter than the pencil and longer than the crayon.

14 The crayon is about _____ units long.

15 The eraser is _____ units shorter than the pen.

Chapter 10 Test A

Test B

Chapter 10 Length

Section A (2 points each)
Circle the correct option: **A, B, C,** or **D.**

1 How many flowers are shorter than Flower K?

K   L   M   N

A 3   **B** 1

C 0   **D** 2

2 Which side is shorter than Side P?

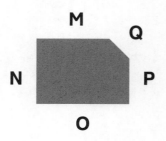

A Q   **B** N

C M   **D** O

3 Use  as 1 unit.

This brush is about _____ units long.

A 2

B 7

C 4

D 9

4 Which ribbon is 7 units longer than Ribbon S?

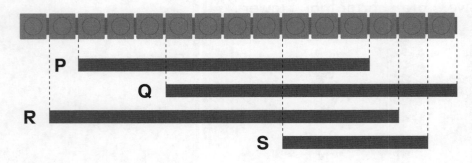

A P

B Q

C R

D S

5 Which one shows the buildings arranged in order, from shortest to tallest?

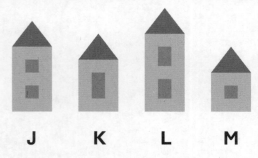

J   K   L   M

A J, K, L, M

B M, K, J, L

C L, J, K, M

D M, L, K, J

Section B (2 points each)

6 Circle the tallest one.

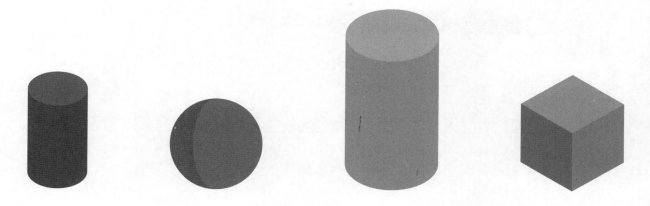

7 Circle the bottle that is as tall as Bottle W.

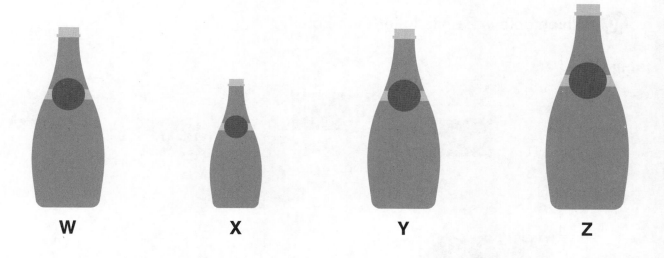

W   X   Y   Z

8 Circle the animal that is taller than the tiger and shorter than the giraffe.

9 Draw a line that is 5 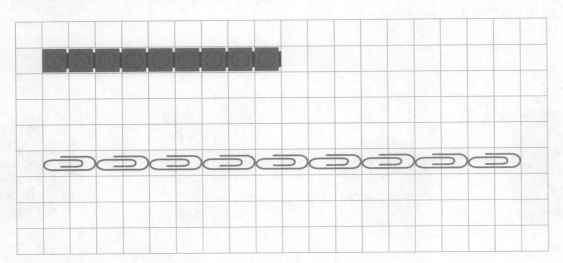 long.
Draw a line that is 5 ⟨⟩ long.

10 Which pole is 2 units taller than Pole C?

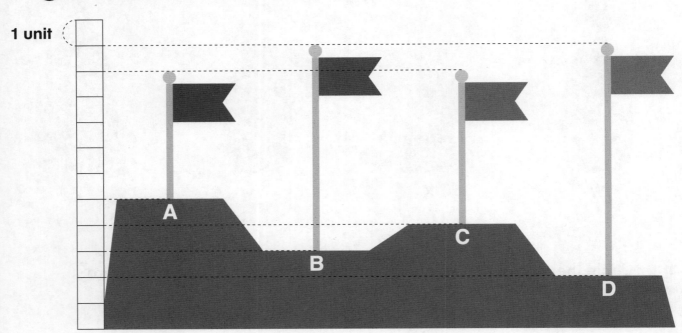

1 unit

Pole _____ is 2 units taller than Pole C.

Look at the picture and answer questions 11 to 13.

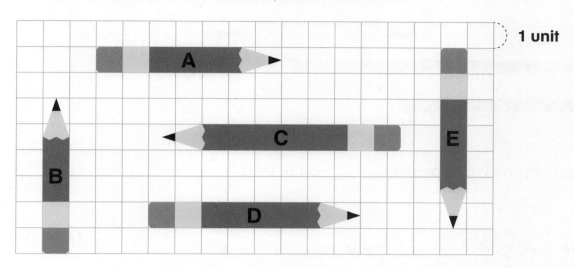

1 unit

11 Pencil D is longer than Pencil A and shorter than Pencil _____.

12 Pencil _____ and Pencil _____ are the same length.

13 Pencil _____ is 3 units shorter than Pencil C.

14 Use  as 1 unit.

Altogether, the two tapes are _____ units long.

15 Grace uses 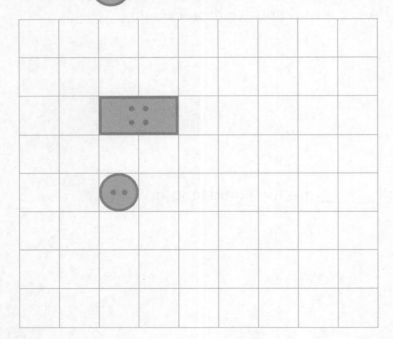 as 1 unit to measure her ribbon.

Siti uses ⬤ as 1 unit to measure her ribbon.

Grace's ribbon is 15 long.

Siti's ribbon is 15 long.

Who has a longer ribbon?

_____ has a longer ribbon.

Test A

Chapter 11 Comparing

Section A (2 points each)
Circle the correct option: **A**, **B**, **C**, or **D**.

1 How many fewer buckets are there than shovels?

A 4   **B** 11

C 0   **D** 7

2 How many more sharpeners are there than erasers?

A 7   **B** 1

C 2   **D** 8

3 There are 8 hamsters.
There are 2 fewer carrots than hamsters.
Which one represents the number of carrots?

A 8 + 2

B 8 − 2

C 6 − 2

D 6 + 2

4 There are 9 nails in the bag.
There are 15 nails in the box.
How many more nails are in the box than in the bag?

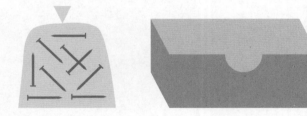

A 9

B 6

C 15

D 3

5 A baker sells 17 donuts and 8 cupcakes.
How many fewer cupcakes than donuts does he sell?

A 8

B 5

C 9

D 20

Section B (2 points each)

There are _____ more than ⬤ .

7 How many fewer shells than starfish are there?

13 − 12 = ☐

There is _____ fewer shell than starfish.

8 Are there more oranges in the bowl or in the bag?
How many more?

☐ − 3 = ☐

There are _____ more oranges in the _____ .

9 There are 5 monkeys and 12 bananas.
How many more bananas than monkeys are there?

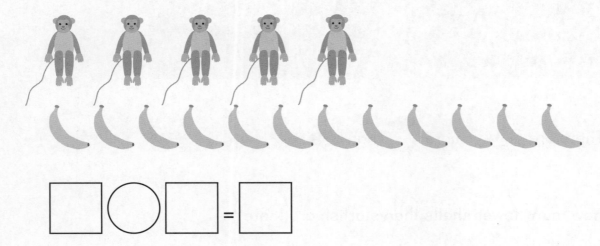

☐ ◯ ☐ = ☐

There are _____ more bananas than monkeys.

10 There are 3 brown eggs in a carton.
There are 9 white eggs in the same carton.

(a) How many more white eggs than brown eggs are there?

☐ ◯ ☐ = ☐

There are _____ more white eggs than brown eggs.

(b) How many eggs are there altogether?

☐ ◯ ☐ = ☐

There are _____ eggs altogether.

This graph shows the number of shells each child has.
Use it to answer questions 11 to 13.

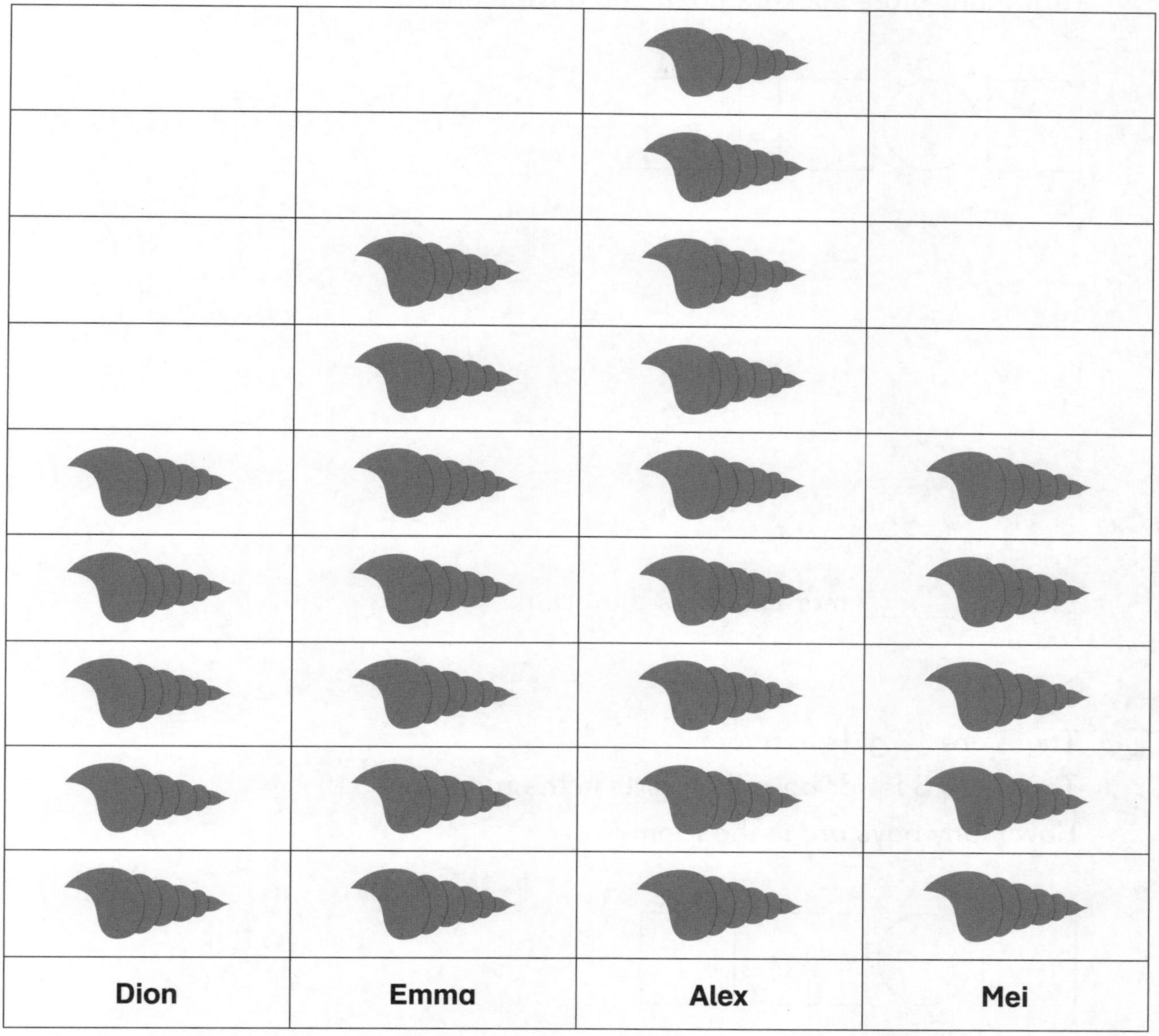

		🐚	
		🐚	
	🐚	🐚	
	🐚	🐚	
🐚	🐚	🐚	🐚
🐚	🐚	🐚	🐚
🐚	🐚	🐚	🐚
🐚	🐚	🐚	🐚
🐚	🐚	🐚	🐚
Dion	**Emma**	**Alex**	**Mei**

11 Alex has _____ shells.

12 _____ has 2 fewer shells than Alex.

13 _____ and _____ have the same number of shells.

14 Dan has 8 stickers.
Ella has 15 stickers.
How many more stickers does Ella have than Dan?

Ella has _____ more stickers than Dan.

15 There are 12 girls in a room.
There are 3 fewer boys than girls in the room.
How many boys are in the room?

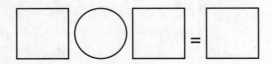

There are _____ boys in the room.

25 min

Score

30

Chapter 11 Comparing

Section A (2 points each)

Circle the correct option: **A**, **B**, **C**, or **D**.

1 There are _____ more 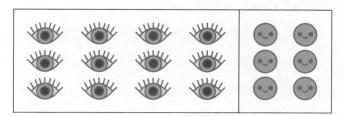 than 😊.

A 6   **B** 7

C 3   **D** 2

2 How many fewer tables than chairs are there?

A 5   **B** 4

C 3   **D** 1

3 There is 1 fewer caterpillar hidden under the leaf than on the stem.
Which one represents the number of caterpillars hidden under the leaf?

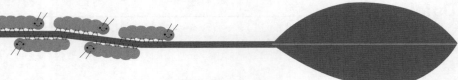

A 5 + 1

B 4 − 1

C 5 − 1

D 5 + 0

4 There are 5 keys outside the bag.
There are 3 more keys inside the bag than there are outside the bag.
Which one represents the number of keys in the bag?

A 5 − 3

B 5 + 3

C 8 − 3

D 8 − 5

5 How many more 🧊 than 🔺 are there?

cone	🔺	🔺	🔺	🔺	🔺	🔺			
cylinder	▬	▬	▬	▬	▬				
cube	🧊	🧊	🧊	🧊	🧊	🧊	🧊	🧊	🧊

A 9

B 1

C 4

D 3

Section B (2 points each)

There are _____ more buttons than needles.

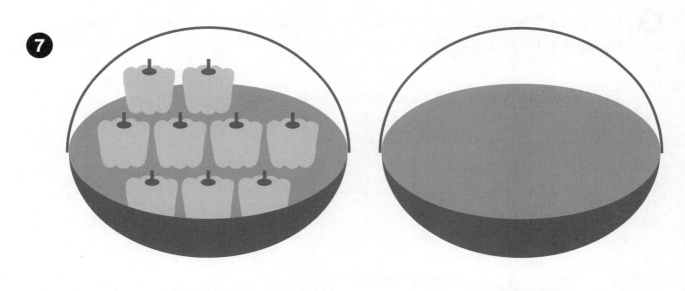

Basket A   **Basket B**

9 – ☐ = ☐

Basket A has _____ more peppers than Basket B.

8 How many more pancakes are on the plate than in the pan?

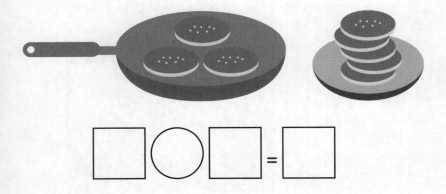

☐ ◯ ☐ = ☐

There are _____ more pancakes on the plate than in the pan.

9 There are 6 fewer pineapples in the bag than there are outside the bag. There are 11 pineapples outside the bag.

(a) How many pineapples are there in the bag?

☐ ◯ ☐ = ☐

There are _____ pineapples in the bag.

(b) How many pineapples are there in all?

☐ ◯ ☐ = ☐

There are _____ pineapples in all.

This graph shows the number of some fruit in a basket.
Use it to answer questions 10 to 12.

			Apple	
			🍎	
			🍎	
			🍎	
	Kiwi		🍎	
	🥝		🍎	
	🥝		🍎	
	🥝		🍎	Strawberry
	🥝		🍎	🍓
Cherry	🥝		🍎	🍓
🍒	🥝		🍎	🍓
🍒	🥝		🍎	🍓
🍒	🥝		🍎	🍓
Cherry	**Kiwi**	**Orange**	**Apple**	**Strawberry**

10 Draw ◯ in the graph to show the same number of oranges as cherries.

11 How many strawberries and kiwis are there altogether?

There are _____ strawberries and kiwis altogether.

12 Are there more apples or kiwis?
How many more?

There are _____ more _____ than _____.

Chapter 11 Test B   23

13 Alex has 8 books.

Emma has 11 books.

Sofia has 5 books.

Who has 3 fewer books than Alex?

_____ has 3 fewer books than Alex.

14 There are 6 red flowers in a vase.

There are 5 fewer red flowers than there are white flowers in the vase.

How many white flowers are in the vase?

☐ ◯ ☐ = ☐

There are _____ white flowers in the vase.

15 Adam has some stickers.

His sister has 4 more stickers than him.

His sister has 9 stickers.

How many stickers does Adam have?

☐ ◯ ☐ = ☐

Adam has _____ stickers.

25 min

Score

30

Test A

Chapter 12 Numbers to 40

Section A (2 points each)
Circle the correct option: **A**, **B**, **C**, or **D**.

1 Twenty-nine is _____.

    A 20   **B** 9

    C 19   **D** 29

2 10 more than 23 is _____.

    A 13   **B** 33

    C 23   **D** 24

3 5 more than 20 is _____.

    A 15   **B** 17

    C 25   **D** 27

4 34 is _____ tens and 4 ones.

A 4   **B** 3

C 7   **D** 10

5 What number is 2 less than 3 tens?

A 23   **B** 10

C 28   **D** 20

6 Complete the number bond.

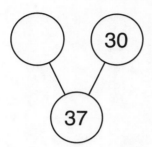

7 Write the missing number.

☐ + 30 = 39

8 Write the missing numbers for this number pattern.

17	19		23		

9 20 less than 32 is _____.

10 Circle the greatest number.

35	29	30	38

11 Circle the numbers in the box that are less than 37 and greater than 27.

25	32	34	26	37	29

12 Write the numbers in the box in order, from least to greatest.

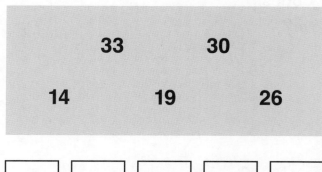

13 Circle the number that shows 1 more than 30.

21   32   29   31   11

14 Carlos has 26 cards.
Adam has 10 fewer cards than Carlos.
How many cards does Adam have?

Adam has _____ cards.

15 The number of beads in a jar is 2 less than 3 tens.
How many beads are in the jar?

_____ beads are in the jar.

25 min **Score**

| 30 |

Chapter 12 Numbers to 40

Section A (2 points each)
Circle the correct option: **A**, **B**, **C**, or **D**.

1 3 tens 7 ones is _____.

   A 30   **B** 37

   C 10   **D** 7

2 0 + 30 is _____.

   A 31   **B** 30

   C 3   **D** 0

3 _____ is 2 less than 26.

   A Twenty-eight   **B** Thirty-four

   C Twenty-two   **D** Twenty-four

4 1 more than 1 ten 5 ones is _____.

A 2 tens 5 ones   **B** 2 tens 6 ones

C 1 ten 1 one   **D** 1 ten 6 ones

5 30 − 1 is _____.

A 20   **B** 29

C 31   **D** 30

6 Complete the number bond.

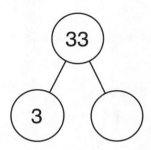

7 Circle the least number and cross out the greatest number.

2 tens 1 one	22	2 ones 1 ten	twenty-eight

8 Write the numbers in the box in order, from greatest to least.

| 20 | 29 | 24 |
| 35 | 30 | 18 |

☐ ☐ ☐ ☐ ☐ ☐

9 Write the missing numbers for this number pattern.

| ☐ | 29 | ☐ | 33 | 35 | ☐ | ☐ |

10 Write the missing number.

37 − 20 = ☐ + 10

11 Circle the numbers in the box that are less than 34 and greater than 17.

15	36	20	33	14	27	37

12 10 less than _____ is 20.

13 What is the same as 1 ten less than 29 ones?

_____ ten _____ ones

14 Amy has 3 sheets of 10 stickers and 5 more stickers.
John has 32 stickers.
Han has thirty-three stickers.
Who has the fewest stickers?

_____ has the fewest stickers.

15 A baker makes 19 cakes.
He makes 20 more donuts than cakes.
How many donuts does he make?

$$\boxed{} + \boxed{} = \boxed{}$$

He makes _____ donuts.

Name: _____

Date: _____

Test A

Chapter 13 Addition and Subtraction Within 40

Section A (2 points each)

Circle the correct option: **A, B, C,** or **D.**

1 23 + 5 = _____

A 18

B 30

C 25

D 28

2 37 – _____ = 35

A 5

B 2

C 7

D 36

3 What is 31 − 3?

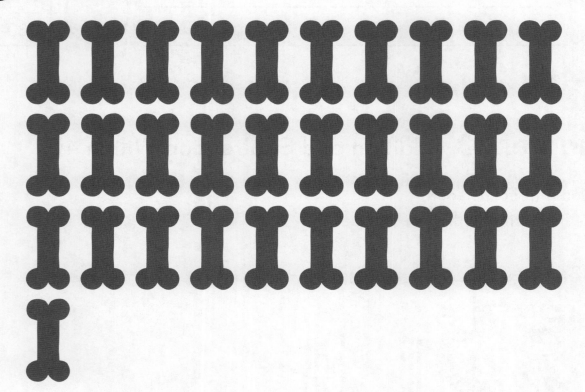

A 28

B 29

C 30

D 34

4 25 + 6 is the same as _____.

A 20 + 16

B 20 + 5

C 20 + 15

D 20 + 11

5 9 + 3 + 6 = _____

A 12

B 15

C 18

D 16

Section B (2 points each)

6 Write the missing number.

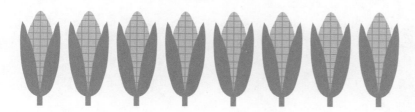

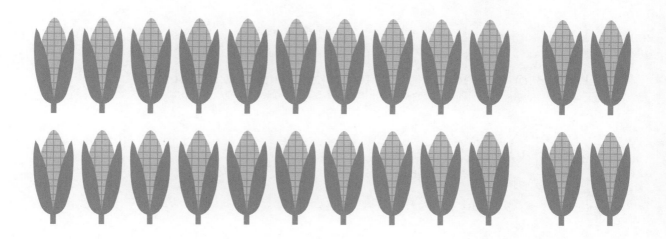

$$8 + 24 = \boxed{}$$

7 Write the missing number.

$$15 + 9 = 10 + \boxed{}$$

8 Write the missing number.

$$34 - 6 = \boxed{}$$

9 Write + or − in each ◯ to make this a true statement.

$$23 \bigcirc 8 = 27 \bigcirc 4$$

10 Write the missing number.

$$2 + 9 + 5 = 9 + 2 + \boxed{}$$

11 There are 5 red crayons, 8 green crayons, and 4 blue crayons in a box.
How many crayons are in the box in all?

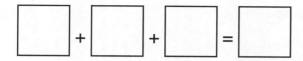

_____ crayons are in the box in all.

12 Kim puts 16 roses and 7 tulips in a vase.
How many flowers are in the vase altogether?

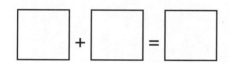

There are _____ flowers in the vase altogether.

13 Laila has 24 stickers.
She gives away 5 stickers.
How many stickers does she have left?

She has _____ stickers left.

14 Alex picks 5 seashells.
Dion picks 8 seashells.
Emma picks 6 seashells.
How many seashells do they pick altogether?

They pick _____ seashells altogether.

15 Sofia has 35 beads.
Mei has 9 fewer beads than Sofia.
How many beads does Mei have?

Mei has _____ beads.

Test B

Chapter 13 Addition and Subtraction Within 40

Section A (2 points each)
Circle the correct option: **A**, **B**, **C**, or **D**.

1 $3 + \underline{\hspace{2cm}} = 37$

   A 7   **B** 34

   C 33   **D** 17

2 $\underline{\hspace{2cm}} = 27 - 6$

   A 26   **B** 7

   C 21   **D** 23

3 5 + 29 = _____ + 30

A 5

B 4

C 10

D 15

4 22 − 7 = 10 + _____

A 12

B 7

C 5

D 17

5 4 + 8 + 7 is the same as _____.

A 8 + 4 + 4

B 7 + 4 + 8

C 4 + 7 + 7

D 8 + 7 + 8

Section B (2 points each)

6 Write the missing number.

$9 + 28 = \boxed{}$

7 Write the missing number.

$\boxed{} = 31 - 9$

8 Write the missing number.

$3 + \boxed{} + 9 = 7 + 9 + 3$

9 Circle the statement that is true.

$17 + 8 = 32 - 6$

$34 - 9 = 6 + 19$

10 Circle the two numbers that together add up to 33.

| 26 | 16 | 6 | 7 | 9 | 25 |

11 There are 36 apples in a basket.
9 of them are green and the rest are red.
How many apples are red?

_____ apples are red.

12 There are 15 children on a bus.
7 more children get on the bus.
How many children are there on the bus now?

There are _____ children on the bus now.

13 Jenna puts 8 coins in an empty jar.
David puts in 7 coins.
Then Sam puts in 3 more coins.
How many coins are in the jar now?

_____ coins are in the jar now.

14 Carter makes 24 jars of jam.

6 of them are apricot jam.

The rest are strawberry jam.

How many jars of strawberry jam does he make?

He makes _____ jars of strawberry jam.

15 Aisha bought 35 cupcakes and 9 donuts for a party.

How many fewer donuts than cupcakes did she buy?

She bought _____ fewer donuts than cupcakes.

25 min   **Score**

30

Test A

Chapter 14 Grouping and Sharing

Section A (2 points each)
Circle the correct option: **A**, **B**, **C**, or **D**.

1 3 fives is _____.

A 5 + 3 + 5

B 3 + 3 + 3

C 5 + 5

D 5 + 5 + 5

2 How many groups of 2 are there?

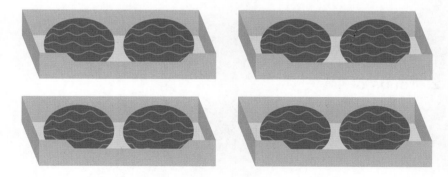

A 4

B 2

C 8

D 1

3 Divide 8 equally into 4 groups.
How many are in each group?

A 4   **B** 2

C 12   **D** 8

4 Pedros has 18 balls.
How many groups of 6 can he make?

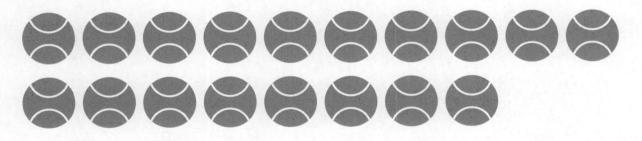

A 3   **B** 6

C 18   **D** 4

5 How many are in 2 groups of 4?

A 2   **B** 16

C 8   **D** 18

Section B (2 points each)

6 Write the missing number.

6 + 6 + 6 = ☐

7 This picture shows _____ groups of 7.

8 Check (✓) the sets that show equal groups.

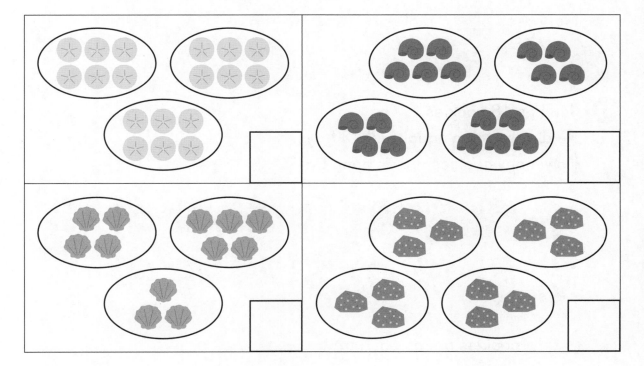

9 Put 16 stars into 2 equal groups.

There are _____ stars in each group.

10 Mary has 8 sandwiches.
She puts 4 sandwiches on each plate.
How many plates does Mary use?

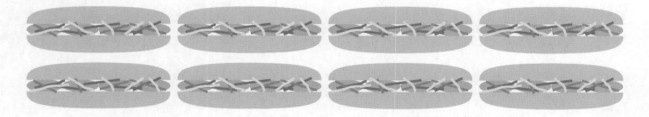

Mary uses _____ plates.

11 Jon has 18 cookies.
He divides them equally into 3 trays.
How many cookies are on each tray?

_____ cookies are on each tray.

12 A baker has 6 cupcakes.

(a) How many groups of 2 can she make?

She can make _____ groups of 2.

(b) How many groups of 3 can she make?

She can make _____ groups of 3.

13 4 children share 12 strawberries equally.
How many strawberries does each child get?

Each child gets _____ strawberries.

14 There are 3 legs on each stool.
How many legs are there on 5 stools?

There are _____ legs on 5 stools.

15 A teacher wants to give each student 2 counters.
There are 10 students.
How many counters does the teacher need?

The teacher needs _____ counters.

25 min   **Score**

30

Test B

Chapter 14 Grouping and Sharing

Section A (2 points each)
Circle the correct option: **A**, **B**, **C**, or **D**.

1 7 + 7 + 7 is the same as _____.

A 2 sevens

B 3 threes

C 3 sevens

D 7 sevens

2 Put 18 pencils equally into 6 groups.
How many pencils are in each group?

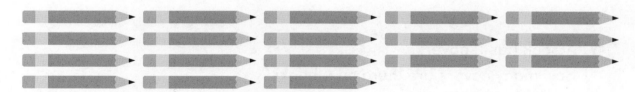

A 6

B 12

C 3

D 4

3 1 + 1 + 1 + 1 + 1 is _____ groups of 1.

A 1   **B** 5

C 11   **D** 0

4 3 groups of 5 is the same as _____.

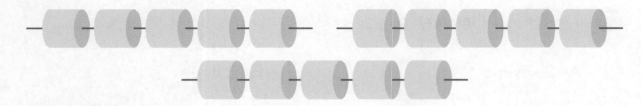

A 3 + 3 + 3   **B** 3 + 5

C 5 + 5 + 5   **D** 5 + 5 + 5 + 5 + 5

5 A book has 2 covers.
How many covers are there on 4 books?

A 10   **B** 7

C 9   **D** 8

Section B (2 points each)

6 Draw  to show 3 threes.

7 Write the missing number.

$$3 + 3 + 3 + 3 + 3 = \boxed{}$$

8 Check (✔) the set that shows 4 groups of 3.

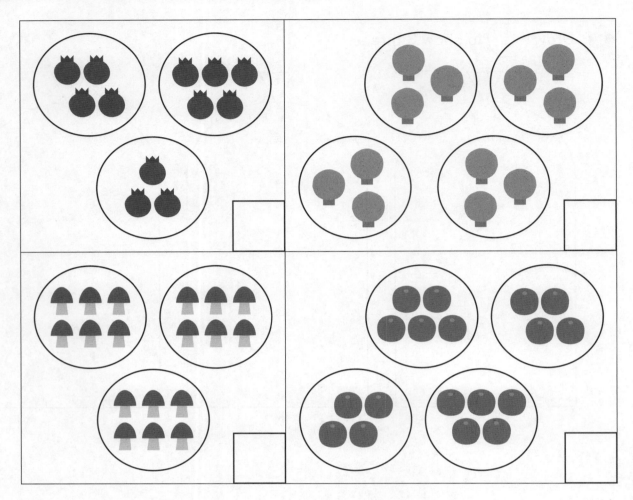

9 There are 15 owls.
Circle groups of 3.

10 Poppy wants to put 4 chocolates in each box.
She has 20 chocolates.
How many boxes does she need?

She needs _____ boxes.

11 John has 14 glue sticks.
He puts them into 2 equal groups.
How many glue sticks are in each group?

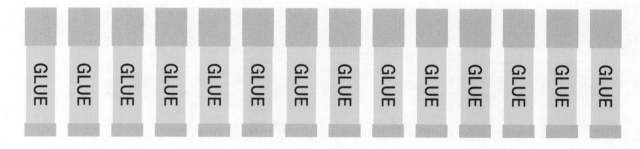

_____ glue sticks are in each group.

12 A teacher gives 18 crayons to some students.
If each student gets 2 crayons, how many students are there?

There are _____ students.

13 There are 6 wheels on each truck.
How many wheels are there on 4 trucks?

There are _____ wheels on 4 trucks.

14 Alex has 2 groups of 6 stickers.
Kona has 3 groups of 3 stickers.
Who has more stickers?

_____ has more stickers.

15 Emma wants to divide 8 books equally into 3 boxes.
How many books will there be in each box?
How many books will be left over?

There will be _____ books in each box.

There will be _____ books left over.

Name: _____

Date: _____

	30

Test A

Chapter 15 Fractions

Section A (2 points each)
Circle the correct option: **A**, **B**, **C**, or **D**.

1 Which one shows a rectangle cut in half?

A

B

C

D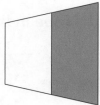

2 Which shape is exactly half shaded?

A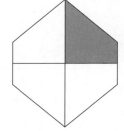

B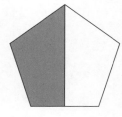

C

D

_____ fourths make 1 whole.

A 4   **B** 1

C 2   **D** 3

4 Which square has been cut into fourths?

A
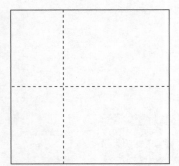

B
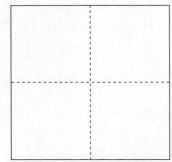

C

D
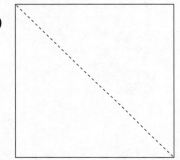

5 What is the next shape in this pattern?

A

B

C

D

Section B (2 points each)

6 Color 1 fourth of this circle.

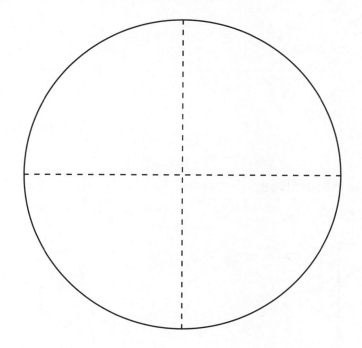

7 Color 1 half of this square.

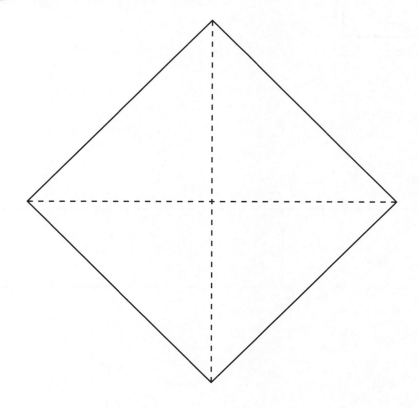

8 Cross out the 2 shapes that are not cut into halves.

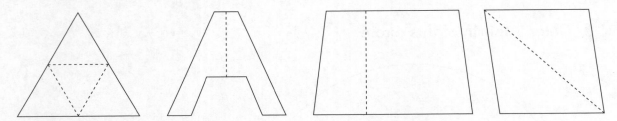

9 Draw a line to cut the square into halves.

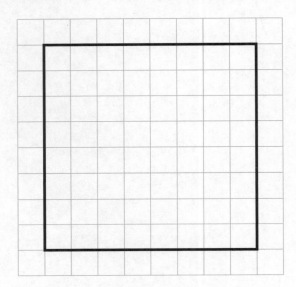

10 Color 1 half of this bar.

Chapter 15 Test A

11 Color the last shape to continue the pattern.

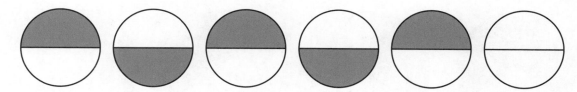

12 Circle the shape that is half shaded.

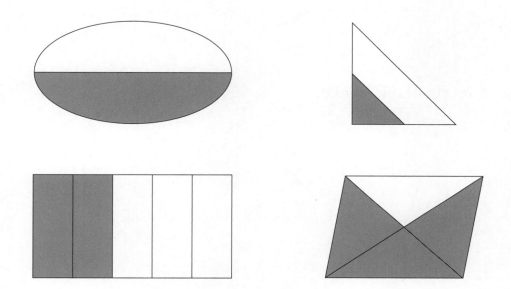

13 Cross out the shape that is one-fourth shaded.

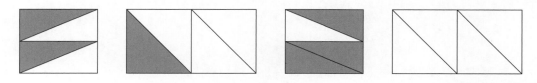

14 Cross out the circle that is half shaded.

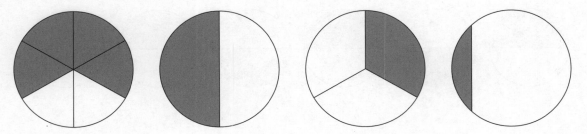

15 Color 1 fourth of this shape.

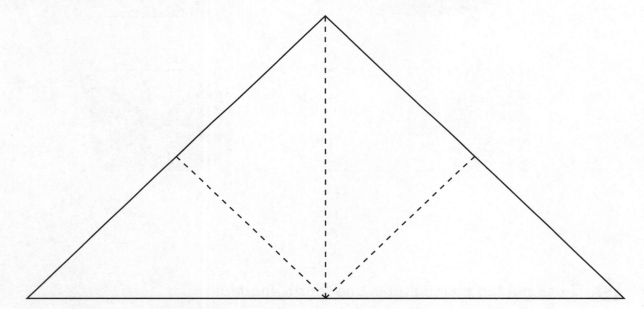

Name: _____

Date: _____

Test B

Chapter 15 Fractions

Section A (2 points each)
Circle the correct option: **A**, **B**, **C**, or **D**.

1 Which shape is cut into halves?

A
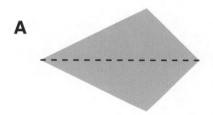

B
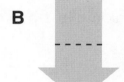

C

D
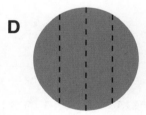

2 Which shape is a quarter circle?

A

B

C

D

3 Which shape is one-fourth shaded?

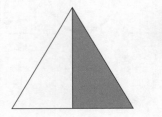 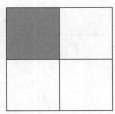 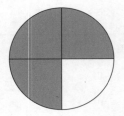 

A circle

B rectangle

C square

D triangle

4 How many fourths make 1 half?

A 1

B 3

C 4

D 2

5 What is the next shape in this pattern?

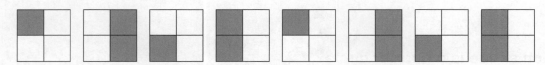

A

B

C

D

Section B (2 points each)

6 Circle the shape that is NOT half shaded.

 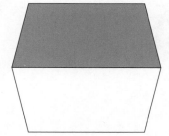

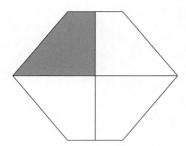 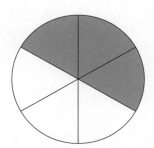

7 Circle the rectangle that is one-fourth shaded.
Cross out the rectangle that is half shaded.

 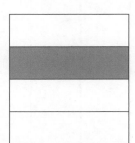

8 Cross out the shape that is one-fourth shaded.

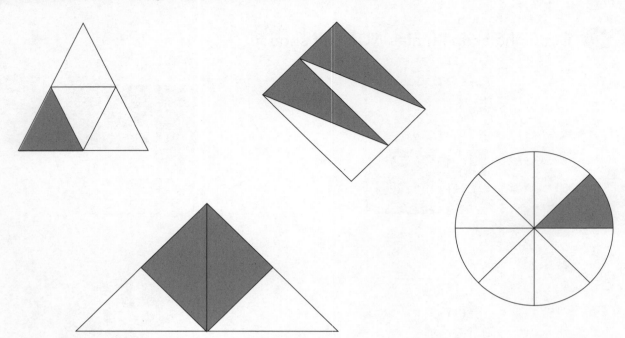

9 Color 1 half of this bar.

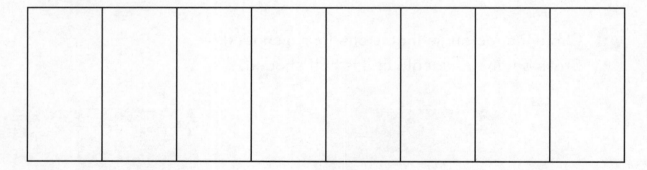

10 Color 1 fourth of this bar.

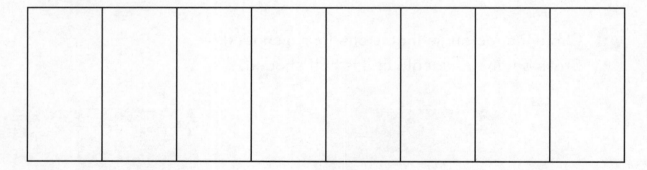

11 Draw 2 lines to cut this shape into fourths.

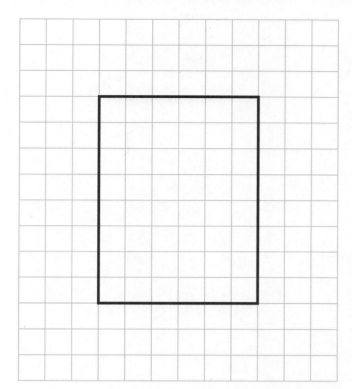

12 Are all the triangles half shaded?
Check (✓) the correct answer.

Yes   No

13 Cross out the shapes that are cut into fourths.

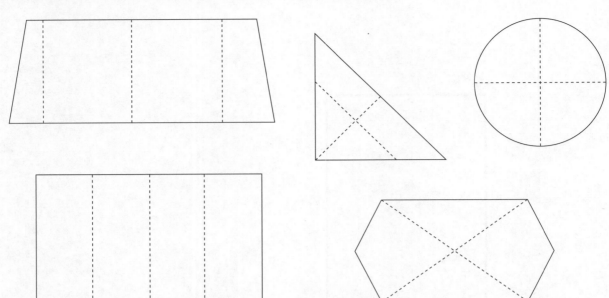

14 Color 1 half of this shape.

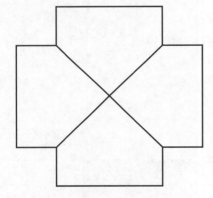

15 I am a square.
One half of me is shaded.
Circle me.

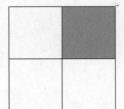 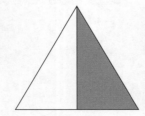  

Test A

Continual Assessment 3

Section A (2 points each)
Circle the correct option: **A**, **B**, **C**, or **D**.

1 Use ⊂⊃ as 1 unit.
How long is the pencil?

A 12 units

B 9 units

C 10 units

D 11 units

2 How many tens and ones are there?

A 3 tens 6 ones

B 2 tens and 7 ones

C 2 tens and 6 ones

D 3 tens and 4 ones

3 29 − 7 is _____.

A 26

B 6

C 22

D 36

4 Which one represents 7 more than 29?

A 27 + 2

B 29 + 7

C 29 − 7

D 20 + 7

5 How many groups of 3 are there?

A 3

B 6

C 9

D 4

6 There are 9 padlocks.
There are 4 fewer keys than padlocks.
Which one shows the number of keys?

A 9 – 4

B 13 – 4

C 13 – 9

D 4 + 9

7 Divide 8 birds equally into 4 nests.
How many birds are in each nest?

A 2

B 8

C 4

D 10

8 7 + 8 + 2 is the same as _____.

A 2 + 7 + 7

B 6 + 10

C 10 + 1

D 8 + 7 + 2

9 3 groups of 8 is the same as _____.

A 3 + 3 + 3

B 8 + 8 + 8

C 3 + 3 + 3 + 3

D 8 + 8 + 8 + 8

10 Which picture shows fourths?

A

B

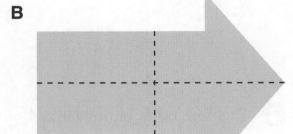

C

D

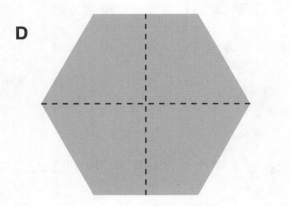

Section B (2 points each)

Look at the picture. Answer questions 11 to 13.

Use ⌒ as 1 unit.

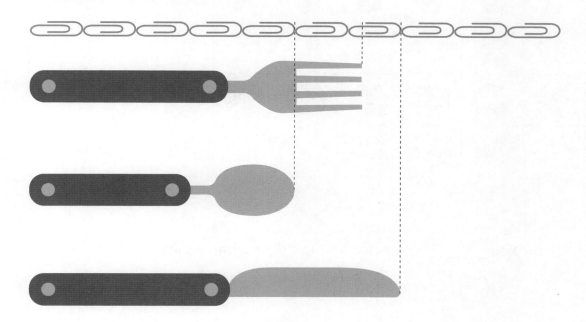

11 The fork is between _____ and _____ units long.

12 The spoon is _____ units shorter than the knife.

13 Altogether, the knife and the spoon are _____ units long.

This graph shows the number of pies a baker makes.
Use it to answer questions 14 to 16.

Cherry	🍒	🍒	🍒	🍒	🍒	🍒	🍒		
Lemon	🍋	🍋	🍋	🍋	🍋				
Apple	🍎	🍎	🍎	🍎	🍎	🍎	🍎	🍎	
Peach	🍑	🍑	🍑	🍑					

14 The baker makes _____ fewer peach pies than cherry pies.

15 If he makes 2 more cherry pies, he will have _____ cherry pies.

16 He makes 13 apple and _____ pies altogether.

17 Arrange the numbers from least to greatest.

25   32   29

19   36

18 Write the missing numbers for the number pattern.

28	30		34		

19 Write the missing number.

$$27 + 9 = 10 + \boxed{}$$

20 Color 1 half of this bar.

Section C (4 points each)

21 There are 28 cars in a parking lot.
6 more come to the parking lot.
How many cars are in the parking lot now?

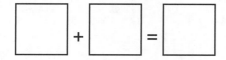

_____ cars are in the parking lot now.

22 Jaime buys 39 eggs.
5 eggs break.
How many eggs are left?

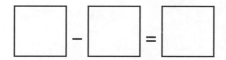

_____ eggs are left.

23 There are 32 books in a bookcase.
There are 20 fewer books on a table than in the bookcase.
How many books are on the table?

_____ books are on the table.

24 Kona put 8 tacos equally on 4 plates.
How many tacos are on each plate?

There are _____ tacos on each plate.

25 Ryan buys 3 bunches of bananas.
There are 5 bananas on each bunch.
How many bananas are there altogether?

There are _____ bananas altogether.

45 min   **Score**

60

Test B

Continual Assessment 3

Section A (2 points each)
Circle the correct option: **A**, **B**, **C**, or **D**.

1 Use ⬭ as 1 unit.

The toothbrush is between _____ units long.

A 7 and 8   **B** 4 and 5

C 5 and 6   **D** 1 and 2

2 Which one is greater than 27 ones and less than 31 ones?

A twenty-four   **B** thirty-three

C thirty   **D** twenty-six

3 35 is _____ more than 3 tens and 3 ones.

A 6

B 2

C 3

D 35

4 _____ = 8 + 29

A 37

B 35

C 38

D 28

5 9 + 5 + 1 is the same as _____.

A 1 + 9 + 9

B 6 + 10

C 9 + 4 + 5

D 8 + 6 + 1

6 Which equation is true?

    A $25 + 6 = 31 - 6$ **B** $30 - 4 = 34 - 8$

    C $25 - 7 = 27 - 8$ **D** $23 + 4 = 27 - 6$

7 A starfish has 5 arms.
There are 15 arms.
How many starfish are there?

    A 5   **B** 20

    C 10   **D** 3

8 An octopus has 8 arms.
There are 3 octopuses.
How many arms are there?

    A 24   **B** 11

    C 16   **D** 8

9 How many beads are needed to make 6 groups of 2?

A 8   **B** 12

C 10   **D** 6

10 Which shape is one-fourth shaded?

A

B

C

D

Look at the picture. Answer questions 11 to 13.

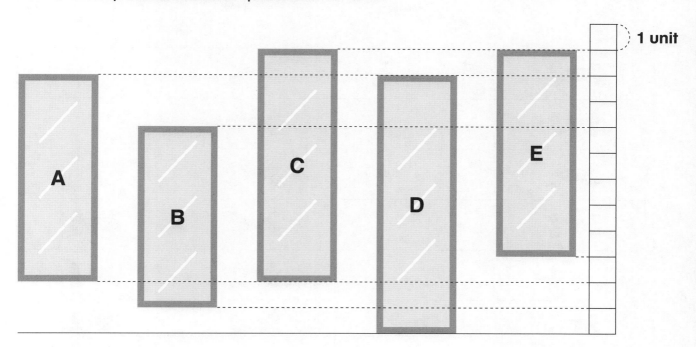

1 unit

11 Mirror B is _____ unit shorter than Mirror E.

12 Mirror _____ is taller than Mirror A and shorter than Mirror D.

13 Altogether, Mirror A and Mirror E are _____ units tall.

This graph shows the number of pets in a pet shop.
Use it to answer questions 14 to 16.

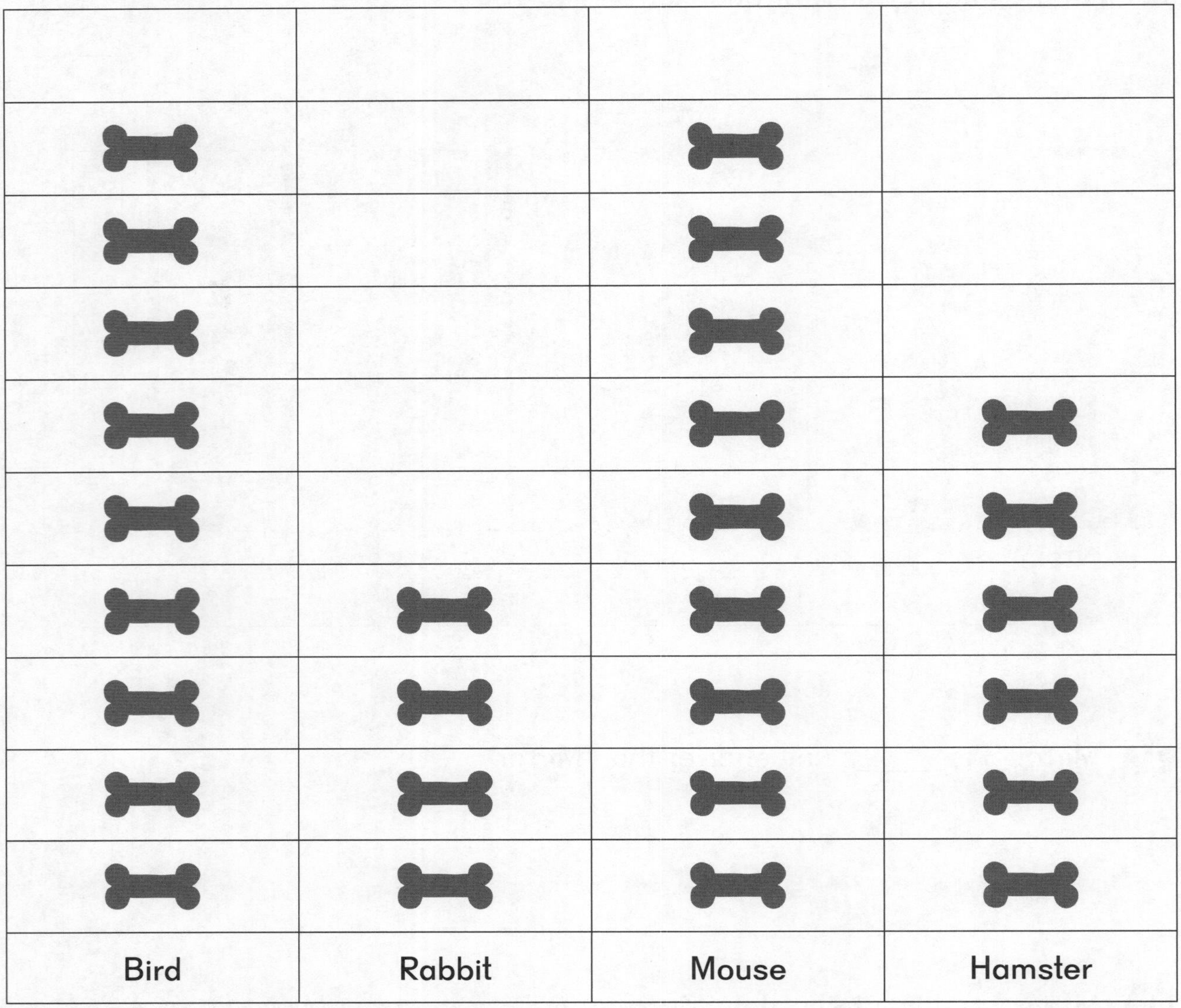

Bird	Rabbit	Mouse	Hamster

14 The shop has the same number of _____ and _____.

15 There are _____ mice, rabbits, and hamsters altogether.

16 The shop sells 6 birds.
There are now _____ fewer birds than hamsters.

17 Circle the two numbers that together add up to 38.

| 28 | 7 | 32 | 9 | 29 | 27 |

18 Write the missing number.

$$17 + \boxed{} = 35 - 9$$

19 Write the missing numbers in the number pattern.

| 33 | 30 | | 24 | | 18 |

20 Color 1 half of this shape.

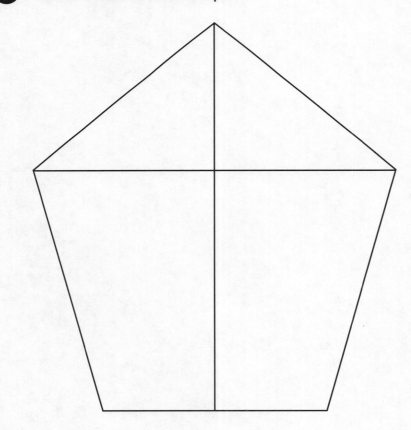

Section C (4 points each)

21 A farmer has 29 apples in a box.
He puts in 6 more apples.
How many apples are in the box now?

There are _____ apples in the box now.

22 Dion has 31 stickers.
He has 8 more stickers than Alex.
How many stickers does Alex have?

Alex has _____ stickers.

23 John has 15 crayons.
He wants to put 3 in each box.
How many boxes does he need?

He needs _____ boxes.

24 Chapa buys 4 cartons of eggs.
There are 6 eggs in each carton.
How many eggs does she buy?

She buys _____ eggs.

25 There are 4 roses, 5 tulips, and 6 daisies in a vase.
How many flowers are in the vase?

There are _____ flowers in the vase.

Name: _____

Date: _____

25 min

Score

30

Test A

Chapter 16 Numbers to 100

Section A (2 points each)

Circle the correct option: **A**, **B**, **C**, or **D**.

1 How many beads are there?

A 52

B 70

C 27

D 72

2 3 less than 64 is _____.

A 63

B 61

C 67

D 34

3 This picture shows _____.

A 3 tens and 5 ones

B 5 tens and 5 ones

C 3 tens and 3 ones

D 5 tens and 3 ones

4 30 less than 100 is _____.

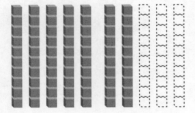

A 97

B 30

C 70

D 100

5 _____ is the same as 2 more than 80.

A 82 + 3

B 3 + 80

C 2 + 82

D 80 + 2

Section B (2 points each)

6 Write the missing numbers.

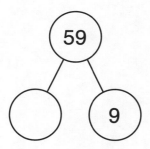

9 less than 59 is ☐.

7 Write the number for eighty-six.

☐

8 Write the missing number.

67 + ☐ = 97

9 _____ is 3 less than 72.

10 Circle the greatest number.

| 58 | 92 | 70 | 89 | 63 |

11 Write the missing numbers for the number pattern.

| 73 | 75 | | 79 | | 83 | |

12 Write the numbers in the box in order, from least to greatest.

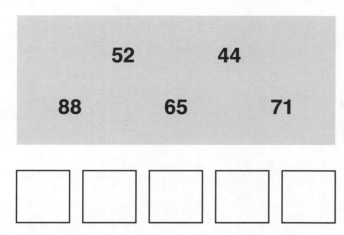

13 Circle the numbers that are less than 6 tens and 2 ones.

63 59 70 65 61

14 Sophia has eighty-five stickers.
Dion has fifty-eight stickers.
Who has fewer stickers?

_____ has fewer stickers.

15 A farmer has 62 white eggs.
He has 3 more brown eggs than white eggs.
How many brown eggs does he have?

He has _____ brown eggs.

Name: _____

Date: _____

25 min

Score

30

Test B

Chapter 16 Numbers to 100

Section A (2 points each)
Circle the correct option: **A**, **B**, **C**, or **D**.

1 40 more than 6 tens is _____.

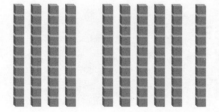

A 100

B 37

C 73

D 70

2 3 less than 81 is _____.

A 80

B 82

C 78

D 84

3 67 is greater than _____.

A 6 tens and 9 ones

B 6 tens

C 7 tens

D 7 tens and 6 ones

4 20 less than 89 is the same as _____.

A 82 − 9

B 69 + 20

C 49 + 20

D 89 − 2

5 76 is between _____.

A 65 and 75

B 60 and 70

C 70 and 74

D 75 and 80

Section B (2 points each)

6 Write the missing numbers.

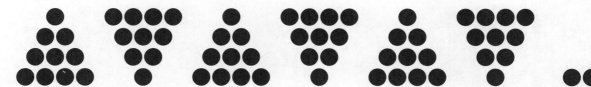

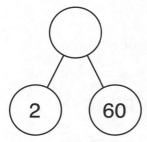

Tens	Ones

7 Write the missing number.

☐ + 69 = 71

8 _____ is greater than 97 and less than 99.

9 Circle the greatest number and cross out the least number.

9 ones and 5 tens	7 tens and 7 ones
10 tens	1 one and 4 tens

10 Write the missing numbers for this number pattern.

87	85			79	77	

11 Write the numbers in the box in order, from greatest to least.

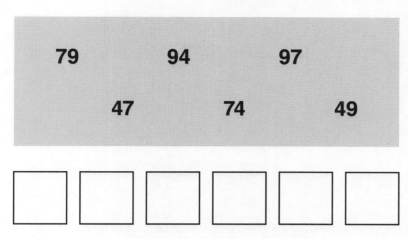

79   94   97

47   74   49

☐ ☐ ☐ ☐ ☐ ☐

12 82 is 6 tens and _____ ones.

13 Circle the numbers in the box that are greater than 44 and less than 57.

| 45 | 38 | 81 | 54 | 43 | 50 | 58 |

14 Mei has 85 beads.
Alex has ninety-two beads.
Dion has 79 beads.
Who has the most beads?

_____ has the most beads.

15 A florist has 52 red roses.
She has 3 fewer white roses than red roses.
How many white roses does the florist have?

The florist has _____ white roses.

25 min

Score

30

Test A

Chapter 17 Addition and Subtraction Within 100

Section A (2 points each)
Circle the correct option: **A**, **B**, **C**, or **D**.

1 What is 50 − 4?

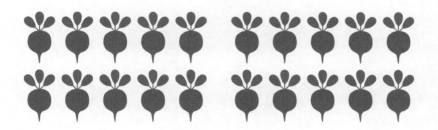

A 44

B 56

C 46

D 14

2 28 + 41 = _____

A 48

B 69

C 42

D 24

3 Subtract 70 from 87.
The answer is _____.

A 10

B 7

C 17

D 80

4 7 tens − 4 tens = _____

A 3

B 74

C 11

D 30

5 95 − 67 = _____

A 25

B 28

C 27

D 26

6 Write the missing number.

$$46 + 21 = \boxed{}$$

7 Write the missing number.

$$65 - 9 = \boxed{} + 5$$

8 Write the missing number.

$$54 + 8 = \boxed{} + 12$$

9 Write + or − in each 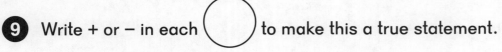 to make this a true statement.

97 ◯ 34 = 41 ◯ 22

10 Fill in the blanks with the numbers that make 77.

| 45 | 58 | 37 | 19 |

_____ and _____ make 77.

11 Mark is 57 years old now.
How old will he be 6 years from now?

He will be _____ years old.

12 Kalama read 25 pages of a book yesterday.
She read 62 pages of the same book today.
How many pages did she read altogether?

She read _____ pages altogether.

13 Jason has 74 stickers.
He uses 11 stickers.
How many stickers does he have left?

He has _____ stickers left.

14 A farmer has 43 white eggs and 18 brown eggs.
How many eggs does she have in all?

She has _____ eggs in all.

15 There are 68 children on a bus.
29 of them brought their own lunch from home.
How many children did not bring their lunch from home?

_____ children did not bring their lunch from home.

25 min   **Score**

30

Test B

Chapter 17 Addition and Subtraction Within 100

Section A (2 points each)
Circle the correct option: **A**, **B**, **C**, or **D**.

1 Add 4 tens to 5 tens and 6 ones.
The answer is _____.

A 11   **B** 56

C 45   **D** 96

2 58 = _____ + 32

A 30   **B** 25

C 26   **D** 22

3 Which statement is true?

A $30 + 42 = 32 + 30$

B $78 - 70 = 63 - 53$

C $25 + 50 = 90 - 15$

D $43 - 12 = 6 + 31$

4 $92 - 7 =$ _____

A 85

B 83

C 72

D 79

5 $62 + 38 =$ _____

A 98

B 90

C 100

D 92

6 Write the missing number.

$$\boxed{} = 49 + 23$$

7 Write the missing number.

$$81 - \boxed{} = 24$$

8 Write the missing number.

$$13 + \boxed{} = 91 - 11$$

9 Make 2 subtraction equations with these numbers.

19   65   46

$\boxed{} - \boxed{} = \boxed{}$

$\boxed{} - \boxed{} = \boxed{}$

10 Circle the two numbers that together add up to 72.

| 18 | 55 | 17 | 56 | 25 | 49 |

11 There are 33 red balloons and 54 yellow balloons at a party.
How many balloons are there in all?

There are _____ balloons in all.

12 A baker made 85 cookies.
He sold 62 of them.
How many cookies were not sold?

_____ cookies were not sold.

13 Emma had 51 jelly beans.
She ate some of them and had 43 left.
How many jelly beans did she eat?

She ate _____ jelly beans.

14 Matt has 16 stickers.

His sister has 76 stickers.

How many stickers do they have altogether?

They have _____ stickers altogether.

15 There are 92 children in the school hall.

47 of them are wearing backpacks.

How many children are not wearing backpacks?

_____ children are not wearing backpacks.

Name: _____

Date: _____

30

Test A

Chapter 18 Time

Section A (2 points each)
Circle the correct option: **A**, **B**, **C**, or **D**.

1 The clock is showing _____.

A 3:30   **B** 8:30

C 9:00   **D** 9:30

2 A quarter past seven is _____.

A 7:00   **B** 7:45

C 7:15   **D** 7:30

3 There are _____ minutes in 1 hour.

A 30

B 100

C 12

D 60

4 The time is _____ minutes to 9.

A 1

B 5

C 8

D 55

5 A bus arrives at a bus stop every 10 minutes.
One just arrived at 12:00.
What time does the next bus arrive?

A 12:20

B 12:10

C 10:00

D 11:50

Chapter 18 Test A

Section B (2 points each)

6 Draw the missing hand.

3:05

7 The clock shows the time now.

In _____ minutes, it will be 5 o'clock.

8 Match.

12:35

half past eight

10 o'clock

9 Write the time.

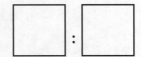

10 Write the time for 5 past 5.

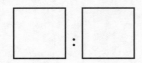

Use the picture to answer questions 11 to 13.

11 The minute hand is pointing to _____.

12 The time is ☐ : ☐ .

13 It is five minutes to _____.

14 Cross out the clock that is showing a little before half past 1.

15 Draw the minute hand.

20 minutes to 3

Name: _____

Date: _____

Score

30

Test B

Chapter 18 Time

Section A (2 points each)
Circle the correct option: **A**, **B**, **C**, or **D**.

1 There are _____ minutes in 1 quarter hour.

A 60

B 15

C 30

D 4

2 The clock shows _____ minutes after 1:45.

A 1

B 15

C 11

D 5

3 A quarter past 10 is the same as _____.

A a quarter to 10

B half past 10

C 15 minutes to 10

D 15 minutes past 10

4 A train leaves from the station every 10 minutes.
One just left at 2:10.
What time does the next train leave?

A 3:00

B 2:30

C 2:20

D 10:10

5 What time is it when both the hour hand and minute hand are pointing to 12?

A 1:00

B 12:00

C 6:00

D 2:00

Section B (2 points each)

6 It is 6:20.

In _____ minutes, it will be 7:00.

7 Cross out the clocks that show the time just before a quarter past 11.

8 Draw the minute hand.

5 minutes after 12

9 The school bus comes at a quarter to 7.
Draw the minute hand to show the time the school bus comes.

Use this picture to answer questions 10 to 12.

10 The time is _____ minutes past 4.

11 The hour hand is between _____ and _____.

12 It is _____ minutes to 5.

13 Dion will be home in 10 minutes at 3:35.
What time is it now?

The time now is ☐ : ☐ .

14 Emma gets on the school bus at 7:20.
Alex gets on the school bus at a quarter past 7.
Who gets on the school bus first?

_____ gets on the school bus first.

15 Mei goes to bed at 8:30.
Sofia goes to bed 20 minutes later than Mei.
What time does Sofia go to bed?

Sofia goes to bed at ☐ : ☐ .

Name: _____

Date: _____

30

Test A

Chapter 19 Money

Section A (2 points each)
Circle the correct option: **A**, **B**, **C**, or **D**.

1 A quarter has the same value as _____.

A 10¢   **B** 5¢

C 20¢   **D** 25¢

2 How much money is shown in the picture?

A 4¢   **B** 45¢

C 50¢   **D** 40¢

3 A 20-dollar bill has the same value as _____ 5-dollar bills.

A 5

B 2

C 4

D 20

4 A 1-dollar bill has the same value as _____ cents.

A 100

B 10

C 1

D 25

5 1 nickel more than 50¢ is _____.

A 51¢

B 60¢

C 15¢

D 55¢

Section B (2 points each)

6 Write the amount of money.

_____ ¢

7 Check (✓) the set that has the least amount of money.

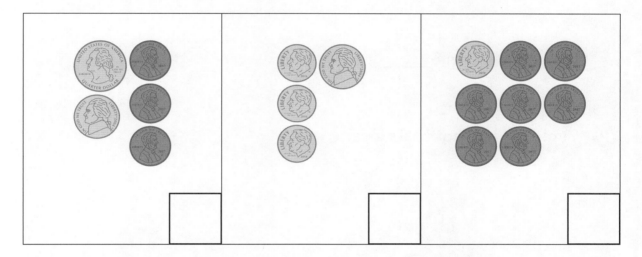

8 Check (✓) the set that has more money.

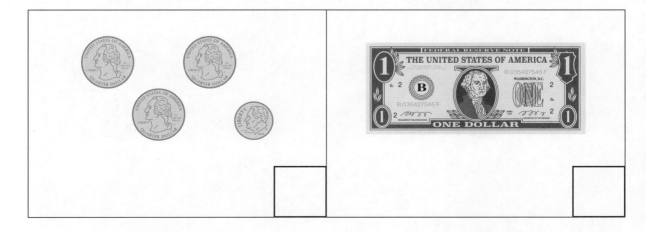

9 How much does the helmet cost?

The helmet costs $ _____.

10 Circle the hat that costs less.

Use the picture to answer questions 11 to 13.

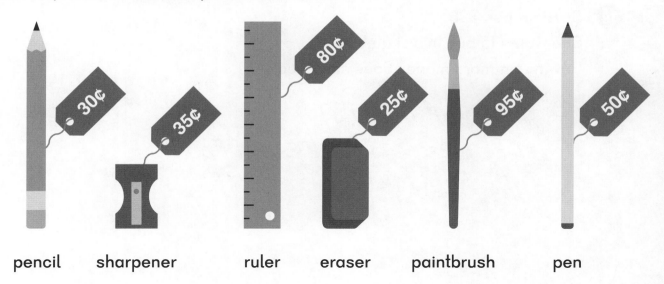

pencil   sharpener   ruler   eraser   paintbrush   pen

11 Alex wants to buy the pen and the sharpener.
How much money does he need?

He needs _____ ¢.

12 Emma has 1 dime, 2 nickels, and 5 pennies.
Which item can she buy?

She can buy the _____.

13 Which item costs 65¢ more than the pencil?

The _____ costs 65¢ more than the pencil.

14 Sushma has $25.
She wants to buy this skateboard.
How much more money does she need?

She needs $ _____ more.

15 Colton paid for a bottle of juice with 4 quarters.
He received 5¢ back.
How much did the bottle of juice cost?

The bottle of juice cost _____ ¢.

Name: _____

Date: _____

30

Test B

Chapter 19 Money

Section A (2 points each)
Circle the correct option: **A**, **B**, **C**, or **D**.

1 2 quarters plus 1 dime is _____ cents.

   A 50   **B** 51

   C 60   **D** 55

2 Jade paid 90¢ for 2 pieces of fruit.
      Which 2 pieces of fruit did she buy?

   A pear and banana   **B** apple and lemon

   C banana and apple   **D** lemon and pear

3 Hunter has these coins.
How much more money does he need to make a dollar?

A 65¢

B 35¢

C 100¢

D 30¢

4 The cake costs _____ more than the pie.

A $34

B $29

C $9

D $16

5 Logan had $7 left after buying this book.
How much money did he have at first?

A $14

B $7

C $21

D $23

Section B (2 points each)

6 Write the amount of money.

_____ ¢

7 How much more money is needed to buy the bracelet?

_____ ¢ more is needed to buy the bracelet.

8 Each bag of crackers costs 25¢.
Jacob has 3 quarters.
How many bags of crackers can he buy?

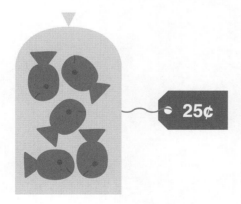

He can buy _____ bags of crackers.

9 Which costs more than the tiger?
How much more?

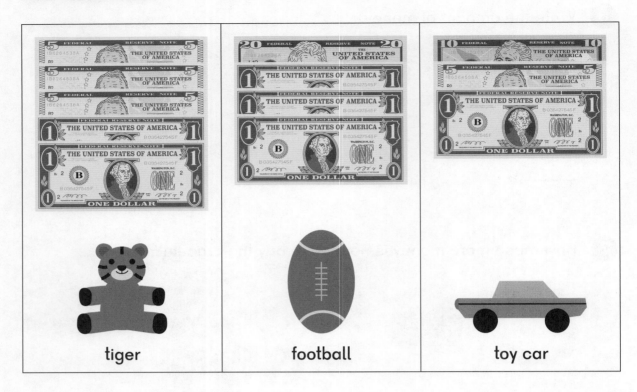

| tiger | football | toy car |

The _____ costs $ _____ more than the tiger.

10 Circle the 3 coins that Alex used to buy this toy.

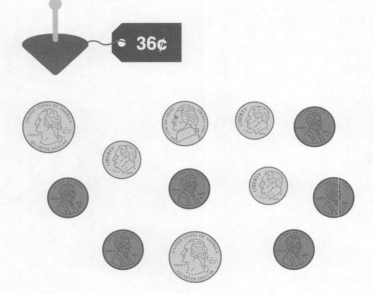

36¢

Use this picture to answer questions 11 to 13.

Dion's Money	Mei's Money	Emma's Money

11 Who has the most money?

_____ has the most money.

12 Who has 52¢ less than Emma?

_____ has 52¢ less than Emma.

13 Emma gives Mei 35¢.
How much money does Mei have now?

Mei has _____ ¢ now.

14 Laila has $25.
She needs $55 more to buy a skateboard.
How much does the skateboard cost?

The skateboard costs $ _____.

15 Carlos bought a drink with a 1-dollar bill.
He received 3 nickels in change.
How much did the drink cost?

The drink cost _____ ¢.

Name: _____

Date: _____

Test A

Year-End Assessment

Section A (2 points each)
Circle the correct option: **A**, **B**, **C**, or **D**.

1 What is the missing shape?

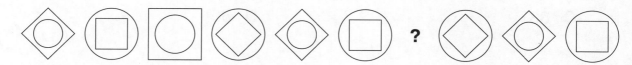

A ◇ (diamond in circle)

B ○ (circle in diamond)

C □ (square in circle)

D ○ (circle in square)

2 Which one has 2 fewer corners than this shape?

A (parallelogram)

B (triangle)

C (square)

D (circle)

3 53 ones is _____.

A 3 tens 5 ones

B 5 tens 3 ones

C 5 tens 5 ones

D 53 tens

4 30 less than 88 is _____.

A 58

B 38

C 83

D 85

5 22 − _____ = 14

A 7

B 6

C 8

D 36

6 Which one is 17 more than 81?

A 87

B 98

C 11

D 91

7 Divide 12 equally into 4 groups.
How many are in each group?

A 3

B 6

C 2

D 4

8 Which shape is one-fourth shaded?

A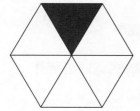

B

C

D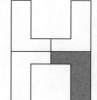

9 78 + 7 is the same as _____.

A 80 + 7

B 80 + 5

C 70 + 5

D 70 + 8

10 Which one shows 4 groups of 1?

A 1 + 4 + 1 + 4

B 5 + 5 + 5 + 5

C 1 + 1 + 1 + 1

D 4 + 4 + 4 + 4

11 Which one shows the tapes arranged from shortest to longest?

A M, N, O, P

B P, M, N, O

C P, O, N, M

D O, N, M, P

12 What is 25 + 60 + 1?

A 86

B 95

C 85

D 81

13 40 − 5 is the same as _____.

A 40 − 35

B 45 − 5

C 42 − 6

D 42 − 7

14 Cody pays for this candy with a 1-dollar bill.
How much change does he get back?

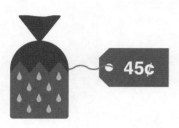

A 45¢

B $55

C 55¢

D $1

15 Mei leaves for school at a quarter past 7.
Which clock shows the time Mei leaves for school?

A

B

C

D

Section B (2 points each)

16 Cross out the shape that does not belong.

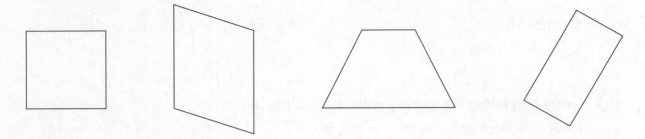

17 Write the missing number.

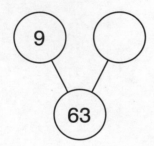

18 Cross out the flamingo that is 3rd from the right.
Circle the flamingo that is in the middle.

left   **right**

This graph shows the items Mary buys at a bakery.
Use it to answer questions 19 to 21.

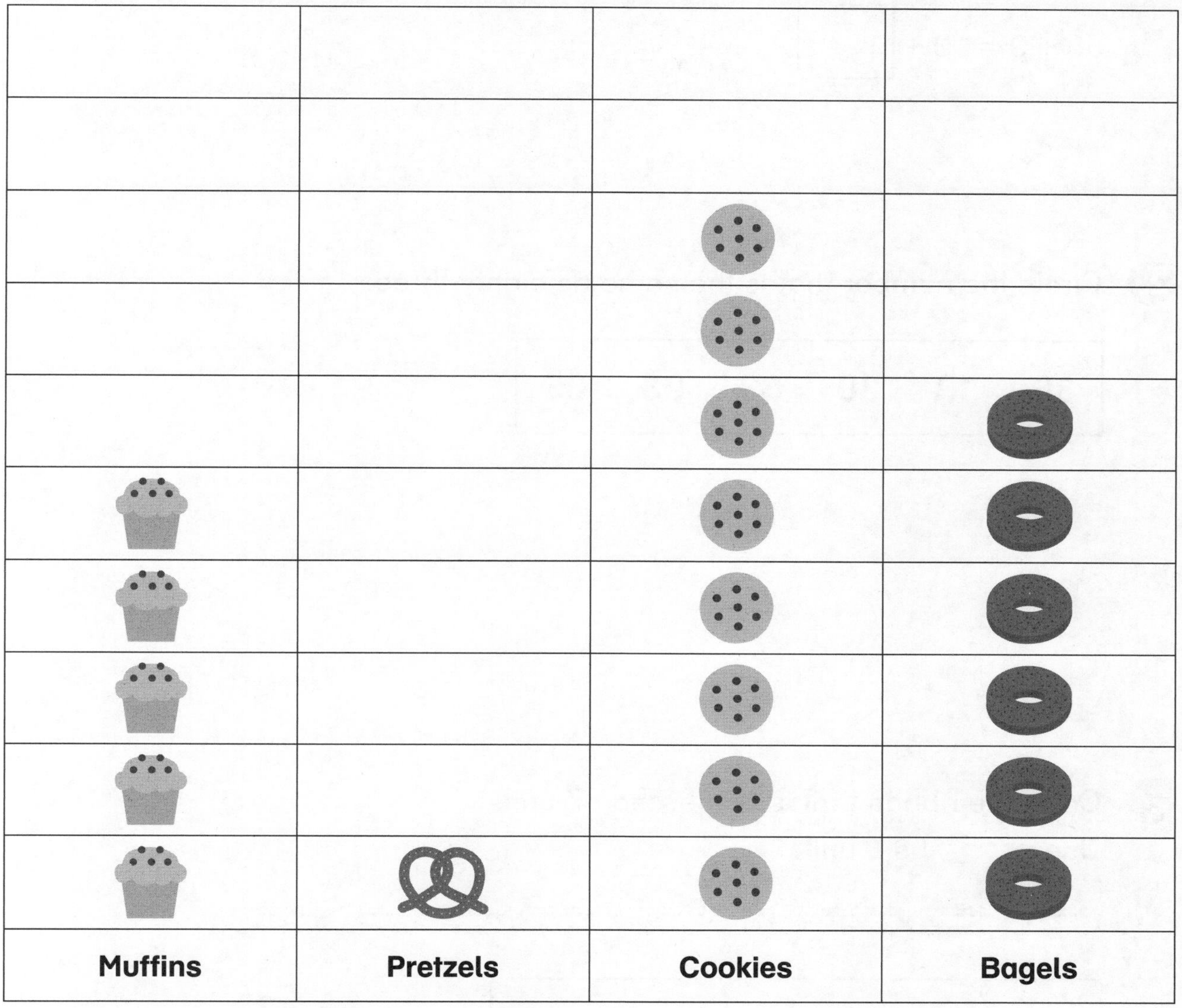

		🍪	
		🍪	
		🍪	🥯
🧁		🍪	🥯
🧁		🍪	🥯
🧁		🍪	🥯
🧁		🍪	🥯
🧁	🥨	🍪	🥯
Muffins	**Pretzels**	**Cookies**	**Bagels**

19 Mary buys 3 fewer _____ than _____.

20 Mary buys _____ muffins, pretzels, and bagels in all.

21 Mary eats 1 pretzel. How many pretzels does she have now?

She has _____ pretzels now.

22 Write the missing number.

$$58 + 9 = 10 + \boxed{}$$

23 Circle the number that is the same as eighty-three.

| 88 | 13 | 38 | 83 | 93 | 35 |

24 Color the ribbon that is longer than 7 units.
Use  as 1 unit.

A	

| **B** |

| **C** |

25 Color 1 half of this shape.

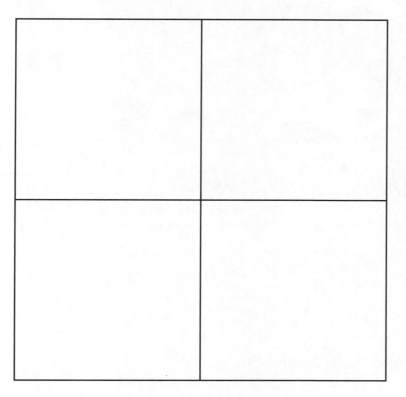

26 Arrange the numbers from least to greatest.

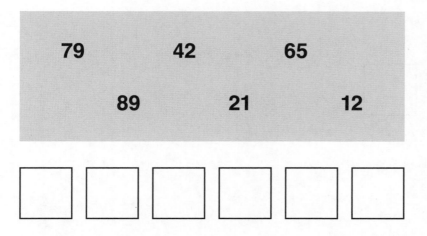

27 Write the missing numbers for the number pattern.

| 71 | | 75 | 77 | | 81 |

28 Write the amount of money.

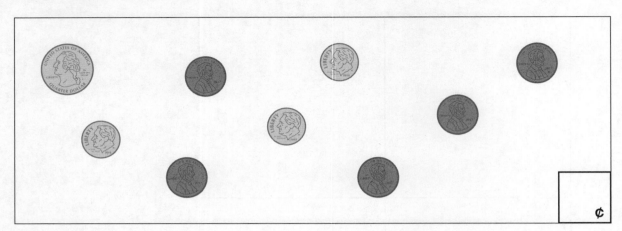

¢

Use the picture to answer questions 29 and 30.

29 The time is ▢ : ▢ .

30 It is ten minutes to _____.

Section C (4 points each)

31 A teacher gave 10 counters to a group of students.
Each student received 2 counters.
How many students did the teacher give counters to?

The teacher gave counters to _____ students.

32 There are 91 red balloons.
There are 3 fewer blue balloons than red balloons.
How many blue balloons are there?

There are _____ blue balloons.

33 There are 33 green pears in a basket.
There are 25 red pears in the same basket.
How many pears are in the basket?

_____ pears are in the basket.

34 A table costs $54.
It costs $34 more than a chair.
How much does the chair cost?

The chair costs $_____.

35 Carlos has 4 boxes of cookies.
There are 6 cookies in each box.
How many cookies does he have?

He has _____ cookies.

60 min   **Score**

80

Test B

Year-End Assessment

Section A (2 points each)
Circle the correct option: **A**, **B**, **C**, or **D**.

1 Grace has 2 pencils that are each 6 ⌒ long.
Which are her two pencils?

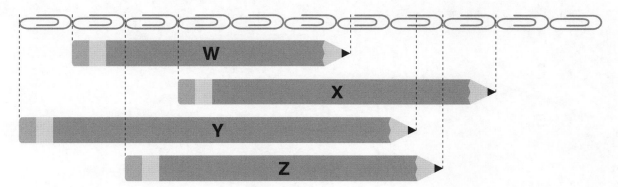

A X and Y   **B** W and Z

C X and Z   **D** W and Y

2 8 ones and _____ tens is 58.

A 50   **B** 5

C 8   **D** 10

3 There are 6 legs on each table.
There 3 tables.
How many table legs are there?

A 9

B 18

C 12

D 6

4 How many corners does the missing shape in this pattern have?

A 3

B 4

C 5

D 0

5 Which one is a triangle with 1 half shaded?

A

B

C

D

6 Which clock shows a time that is a little after 10:50?

A

B

C

D

7 57 + 27 = _____

   A 84   **B** 77

   C 87   **D** 74

8 98 − _____ = 5

   A 88   **B** 95

   C 90   **D** 93

9 Nora pays 2 quarters, a nickel, and 2 dimes for a snack. How much money is the snack?

   A 50¢   **B** 55¢

   C 75¢   **D** 80¢

10 Jordan wants to put 3 oranges into each bag.
He has 18 oranges.
How many bags does he need?

A 9

B 6

C 21

D 18

11 John counts 20 legs on some horses.
How many horses are there?

A 12

B 10

C 5

D 4

12 Which one gives the least number?

A 74 − 51

B 28 + 5

C 1 + 39

D 98 − 84

13 Sofia has 3 coins.
She has 40¢.
What are her 3 coins?

A 1 penny, 1 dime, 1 quarter

B 1 quarter, 1 nickel, 1 dime

C 1 dime, 1 penny, 1 nickel

D 1 quarter, 1 nickel, 1 penny

14 Which shape has 3 corners and 3 sides?

A

B

C

D

15 Dion gets on the school bus at 7:00.
Sofia gets on the bus at 7:30.
Alex gets on the bus after Dion but before Sofia.
What time does Alex get on the bus?

A 7:35

B 7:15

C 7:30

D 6:55

Section B (2 points each)

16 Cross out the shapes that are between the two cones.
Circle the shape that is 2nd from left.

left   **right**

17 Arrange the numbers from least to greatest.

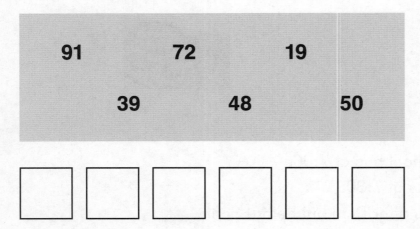

This graph shows the number of crayons Dion has in a box.
Use it to answer questions 18 to 20.

Red	X	X	X	X	X	X	X	X	X	X	X	X
Blue	X	X	X	X	X	X	X	X				
Yellow	X	X	X	X	X	X	X	X	X	X		
Green	X	X	X	X								
Purple												

18 Dion needs to put _____ more green crayons into the box to have the same number of green crayons as red crayons.

19 There are _____ purple crayons in the box.

20 There are _____ red, blue, and yellow crayons in the box altogether.

21 Write the missing numbers.

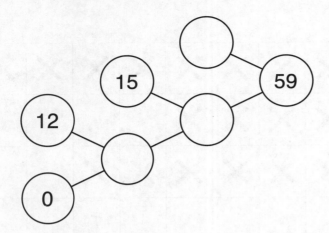

22 Color 1 fourth of this shape.

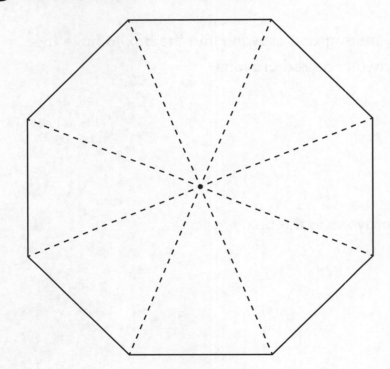

23 Write + or − in each ◯ to make the statement true.

62 ◯ 28 = 50 ◯ 16

Look at the picture. Answer questions 24 and 25.

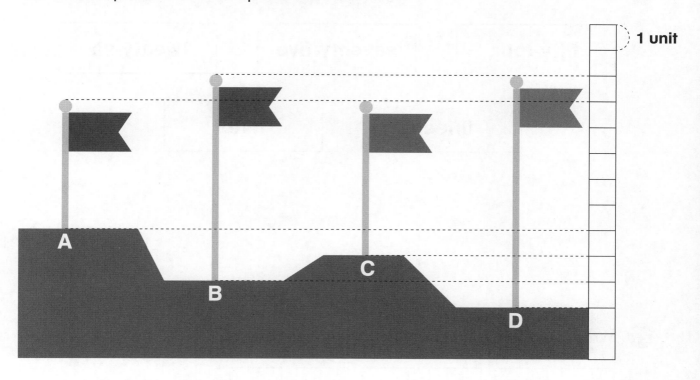

1 unit

24 Pole A is _____ units shorter than Pole D.

25 Write the flag poles in order, from tallest to shortest.

_____ _____ _____ _____

26 Circle the 2 numbers that together make 84.

| fifty-four | seventy-five | twenty-six |

| fifteen | nine |

27 Write the missing numbers in the number pattern.

| | 89 | 79 | | | 49 |

28 Write the missing number.

$$26 + \boxed{} = 73 - 8$$

29 Laila pays two 20-dollar bills and one 10-dollar bill for a pair of shoes.
She receives $8 back.
How much did the pair of shoes cost?

The pair of shoes cost $_____.

30 Check (✓) the statement that is NOT true.

There are 4 quarter hours in an hour.	
There are 3 half hours in an hour.	
There are 15 minutes in a quarter hour.	

31 Emma and Dion are in a line.
Emma is 3rd in line and Dion is last in line.
There are 4 people between Emma and Dion.
How many people are in line?

_____ people are in line.

32 A farmer has 98 red apples.
He has 31 fewer green apples than red apples.
How many green apples does he have?

The farmer has _____ green apples.

33 Ella has $51.
She has $13 more than Joe.
How much money does Joe have?

Joe has $_____.

34 Mei went to bed at 8:00.
Dion went to bed half an hour after Mei.
Emma went to bed 15 minutes after Dion.
What time did Emma go to bed?

Emma went to bed at ☐ : ☐ .

35 Landon wants to put 4 burritos on each plate.
He has 16 burritos.
How many plates does he need?

He needs _____ plates.

Answer Key

Chapter 10 Length

1 C

2 D

3 B

4 B

5 D

6 Side <u>C</u> is the shortest side.

7
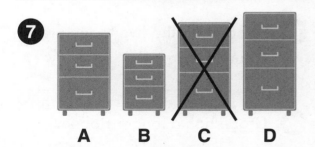
A   B   C   D

8

9 Tape <u>D</u> is longer than Tape A and shorter than Tape C.

10 Answers may vary.

11 Tree <u>B</u> and Tree <u>D</u> have the same height.

12 <u>C</u> <u>A</u> <u>B</u>

13 The <u>pen</u> is shorter than the pencil and longer than the crayon.

14 The crayon is about <u>5</u> units long.

15 The eraser is <u>5</u> units shorter than the pen.

Chapter 10 Length

1 B

---

2 A

---

3 B

---

4 C

---

5 B

6

7

W   X   Y   Z

8

9

---

10 Pole <u>B</u> is 2 units taller than Pole C.

---

11 Pencil D is longer than Pencil A and shorter than Pencil <u>C</u>.

---

12 Pencil <u>A</u> and Pencil <u>E</u> are the same length.

---

13 Pencil <u>B</u> is 3 units shorter than Pencil C.

---

14 Altogether, the two tapes are <u>10</u> units long.

---

15 <u>Grace</u> has a longer ribbon.

Chapter 11 Comparing

1 D

2 B

3 B

4 B

5 C

6 There are <u>3</u> more than .

7 $13 - 12 = \boxed{1}$

There is <u>1</u> fewer shell than starfish.

8 $\boxed{5} - 3 = \boxed{2}$

There are <u>2</u> more oranges in the <u>bowl</u>.

9 $\boxed{12} \bigcirc - \boxed{5} = \boxed{7}$

There are <u>7</u> more bananas than monkeys.

10 (a) $\boxed{9} \bigcirc - \boxed{3} = \boxed{6}$

There are <u>6</u> more white eggs than brown eggs.

(b) $\boxed{3} \bigcirc + \boxed{9} = \boxed{12}$

There are <u>12</u> eggs together.

11 Alex has <u>9</u> shells.

12 <u>Emma</u> has 2 fewer shells than Alex.

13 <u>Dion</u> and <u>Mei</u> have the same number of shells.

14 $\boxed{15} \bigcirc - \boxed{8} = \boxed{7}$

Ella has <u>7</u> more stickers than Dan.

15 $\boxed{12} \bigcirc - \boxed{3} = \boxed{9}$

There are <u>9</u> boys in the room.

Chapter 11 Comparing

1 A

2 C

3 C

4 B

5 D

6 There are <u>2</u> more buttons than needles.

7 $9 - \boxed{0} = \boxed{9}$

Basket A has <u>9</u> more peppers than Basket B.

8 $\boxed{5} - \boxed{3} = \boxed{2}$

There are <u>2</u> more pancakes on the plate than in the pan.

9 (a) $\boxed{11} - \boxed{6} = \boxed{5}$

There are <u>5</u> pineapples in the bag.

(b) $\boxed{11} + \boxed{5} = \boxed{16}$

There are <u>16</u> pineapples in all.

10

Cherry	Kiwi	Orange	Apple	Strawberry

11 There are <u>13</u> strawberries and kiwis altogether.

12 There are <u>3</u> more <u>apples</u> than <u>kiwis</u>.

13 <u>Sofia</u> has 3 fewer books than Alex.

14 $\boxed{6} + \boxed{5} = \boxed{11}$

There are <u>11</u> white flowers in the vase.

15 $\boxed{9} - \boxed{4} = \boxed{5}$

Adam has <u>5</u> stickers.

Chapter 12 Numbers to 40

1 D

2 B

3 C

4 B

5 C

6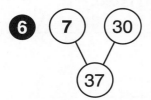

7 | 9 | + 30 = 39

8 17 19 **21** 23 **25** **27**

9 20 less than 32 is <u>12</u>.

10 35 29 30 (38)

11 25 (32)(34) 26 37 (29)

12 14 19 26 30 33

13 21 32 29 (31) 11

14 | 26 | – | 10 | = | 16 |

Adam has <u>16</u> cards.

15 <u>28</u> beads are in the jar.

Chapter 12 Numbers to 40

1 B

_____

2 B

_____

3 D

_____

4 D

_____

5 B

_____

6

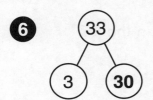

7

2 tens 1 one	22
2 ones 1 ten	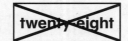

8

35	30	29	24

20	18

9

27	29	**31**	33

35	**37**	**39**

_____

10 37 − 20 = [7] + 10

_____

11 15 36 (20)(33) 14 (27) 37

_____

12 10 less than <u>30</u> is 20.

_____

13 <u>1</u> ten <u>9</u> ones

_____

14 <u>John</u> has the fewest stickers.

_____

15 [19] + [20] = [39]

He makes <u>39</u> donuts.

Chapter 13 Addition and Subtraction Within 40

1 D

2 B

3 A

4 D

5 C

6 $8 + 24 = \boxed{32}$

7 $15 + 9 = 10 + \boxed{14}$

8 $34 - 6 = \boxed{28}$

9 $23 \boxed{+} 8 = 27 \boxed{+} 4$

10 $2 + 9 + 5 = 9 + 2 + \boxed{5}$

11 $\boxed{5} + \boxed{8} + \boxed{4} = \boxed{17}$

17 crayons are in the box in all.

12 $\boxed{16} + \boxed{7} = \boxed{23}$

There are 23 flowers in the vase altogether.

13 She has 19 stickers left.

14 They pick 19 seashells altogether.

15 Mei has 26 beads.

Chapter 13 Addition and Subtraction Within 40

1 B

2 C

3 B

4 C

5 B

6 9 + 28 = | 37 |

7 | 22 | = 31 − 9

8 3 + | 7 | + 9 = 7 + 9 + 3

9 17 + 8 = 32 − 6

(34 − 9 = 6 + 19)

10 | (26) 16 6 (7) 9 25 |

11 | 36 | − | 9 | = | 27 |

<u>27</u> apples are red.

12 | 15 | (+) | 7 | = | 22 |

There are <u>22</u> children on the bus now.

13 <u>18</u> coins are in the jar now.

14 He makes <u>18</u> jars of strawberry jam.

15 She bought <u>26</u> fewer donuts than cupcakes.

Chapter 14 Grouping and Sharing

1 D

2 A

3 B

4 A

5 C

6 6 + 6 + 6 = | 18 |

7 This picture shows <u>3</u> groups of 7.

8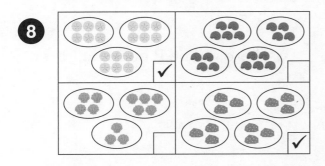

9 There are <u>8</u> stars in each group.

10 Mary uses <u>2</u> plates.

11 <u>6</u> cookies are on each tray.

12 (a) She can make <u>3</u> groups of 2.

(b) She can make <u>2</u> groups of 3.

13 Each child gets <u>3</u> strawberries.

14 There are <u>15</u> legs on 5 stools.

15 The teacher needs <u>20</u> counters.

Chapter 14 Grouping and Sharing

1 C

2 C

3 B

4 C

5 D

6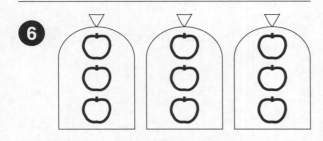

7 3 + 3 + 3 + 3 + 3 = 15

8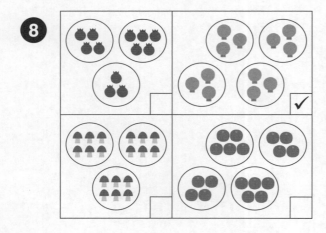

9

10 She needs 5 boxes.

11 7 glue sticks are in each group.

12 There are 9 students.

13 There are 24 wheels on 4 trucks.

14 Alex has more stickers.

15 There will be 2 books in each box.

There will be 2 books left over.

Chapter 15 Fractions

1 C

---

2 D

---

3 A

---

4 B

---

5 A

6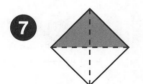

Answers may vary.

7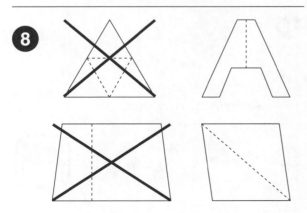

Answers may vary.

8

9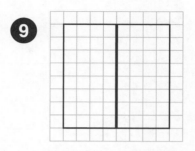

Answers may vary.

10

Answers may vary.

11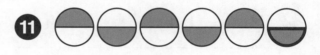

12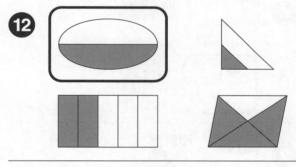

13

14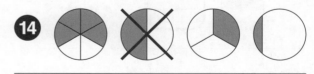

15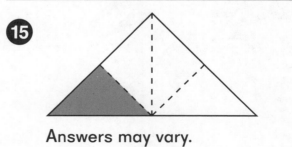

Answers may vary.

Chapter 15 Fractions

1 A

2 C

3 C

4 D

5 C

6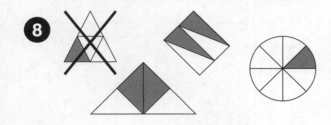

7

8

9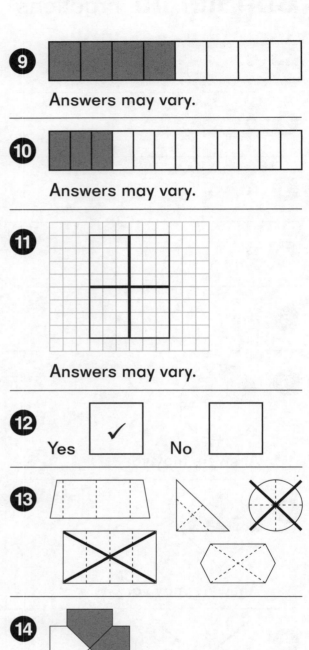

Answers may vary.

10 Answers may vary.

11 Answers may vary.

12 Yes ✓ No

13

14

Answers may vary.

15

Continual Assessment 3

1 B

2 A

3 C

4 B

5 A

6 A

7 A

8 D

9 B

10 D

11 The fork is between <u>6</u> and <u>7</u> units long.

12 The spoon is <u>2</u> units shorter than the knife.

13 Altogether, the knife and the spoon are <u>12</u> units long.

14 The baker makes <u>3</u> fewer peach pies than cherry pies.

15 If he makes 2 more cherry pies, he will have <u>9</u> cherry pies.

16 He makes 13 apple and <u>lemon</u> pies altogether.

17

| 19 | 25 | 29 | 32 |

| 36 |

18

| 28 | 30 | 32 | 34 |

| 36 | 38 |

19 $27 + 9 = 10 + \boxed{26}$

Continual Assessment 3

20

Answers may vary.

21 28 + 6 = 34

<u>34</u> cars are in the parking lot now.

22 39 − 5 = 34

<u>34</u> eggs are left.

23 32 (−) 20 = 12

<u>12</u> books are on the table.

24 There are <u>2</u> tacos on each plate.

25 There are <u>15</u> bananas altogether.

Continual Assessment 3

1 B

2 C

3 B

4 A

5 D

6 B

7 D

8 A

9 B

10 D

11 Mirror B is <u>1</u> unit shorter than Mirror E.

12 Mirror <u>C</u> is taller than Mirror A and shorter than Mirror D.

13 Altogether, Mirror A and Mirror E are <u>16</u> units tall.

14 The shop has the same number of <u>birds</u> and <u>mice</u>.

15 There are <u>19</u> mice, rabbits, and hamsters altogether.

16 There are now <u>3</u> fewer birds than hamsters.

17 28   7   32   (9) (29) 27

18 17 + 9 = 35 − 9

19 33   30   27   24

21   18

Continual Assessment 3

20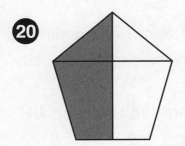

Answers may vary.

21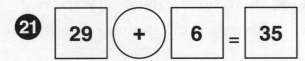

There are <u>35</u> apples in the box now.

22

| 31 | (–) | 8 | = | 23 |

Alex has <u>23</u> stickers.

23 He needs <u>5</u> boxes.

24 She buys <u>24</u> eggs.

25 There are <u>15</u> flowers in the vase.

Chapter 16 Numbers to 100

1 D

2 B

3 D

4 C

5 D

6
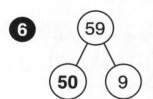

9 less than 59 is 50 .

7 86

8 67 + 30 = 97

9 69 is 3 less than 72.

10 58 92 70 89 63

11 73 75 77 79

 81 83 85

12 44 52 65 71

 88

13 63 59 70 65 61

14 Dion has fewer stickers.

15 He has 65 brown eggs.

Chapter 16 Numbers to 100

1 A

2 C

3 B

4 C

5 D

6

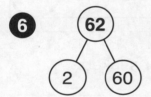

Tens	Ones
6	2

7 [2] + 69 = 71

8 <u>98</u> is greater than 97 and less than 99.

9

9 ones and 5 tens	7 tens and 7 ones
10 tens	~~1 one and 4 tens~~

10

87	85	83	81

79	77	75

11

97	94	79	74

49	47

12 82 is 6 tens and <u>22</u> ones.

13 (45) 38 81 (54) 43 (50) 58

14 <u>Alex</u> has the most beads.

15 The florist has <u>49</u> white roses.

Chapter 17 Addition and Subtraction Within 100

1 C

2 B

3 C

4 D

5 B

6 $46 + 21 = \boxed{67}$

7 $65 - 9 = \boxed{51} + 5$

8 $54 + 8 = \boxed{50} + 12$

9 $97 \bigcirc{-} 34 = 41 \bigcirc{+} 22$

10 <u>58</u> and <u>19</u> make 77.

11 He will be <u>63</u> years old.

12 She read <u>87</u> pages altogether.

13 He has <u>63</u> stickers left.

14 She has <u>61</u> eggs in all.

15 <u>39</u> children did not bring their lunch from home.

Chapter 17 Addition and Subtraction Within 100

1 D

2 C

3 C

4 A

5 C

6 $\boxed{72} = 49 + 23$

7 $81 - \boxed{57} = 24$

8 $13 + \boxed{67} = 91 - 11$

9 $\boxed{65} - \boxed{19} = \boxed{46}$

$\boxed{65} - \boxed{46} = \boxed{19}$

10 $\boxed{18 \quad (55) \quad (17) \quad 56 \quad 25 \quad 49}$

11 There are <u>87</u> balloons in all.

12 <u>23</u> cookies were not sold.

13 She ate <u>8</u> jelly beans.

14 They have <u>92</u> stickers altogether.

15 <u>45</u> children are not wearing backpacks in the school hall.

Chapter 18 Time

1 D

2 C

3 D

4 B

5 B

6

7 In <u>20</u> minutes, it will be 5 o'clock.

8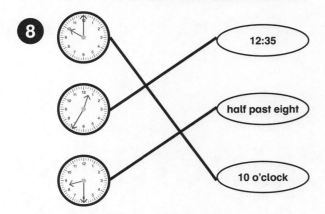

9 11 : 25

10 5 : 05

11 The minute hand is pointing to <u>11</u>.

12 The time is 5 : 55 .

13 It is five minutes to <u>6</u>.

14

15

Chapter 18 Time

1 B

2 D

3 D

4 C

5 B

6 In <u>40</u> minutes, it will be 7:00.

7

8

9

10 The time is <u>35</u> minutes past 4.

11 The hour hand is between <u>4</u> and <u>5</u>.

12 It is <u>25</u> minutes to 5.

13 The time now is $\boxed{3}$: $\boxed{25}$.

14 <u>Alex</u> gets on the school bus first.

15 Sofia goes to bed at

$\boxed{8}$: $\boxed{50}$.

Chapter 19 Money

1 D

2 B

3 C

4 A

5 D

6 57¢

7

8

9 The helmet costs $27.

10

11 He needs 85¢.

12 She can buy the eraser.

13 The paintbrush costs 65¢ more than the pencil.

14 She needs $23 more.

15 The bottle of juice cost 95¢.

Chapter 19 Money

1 C

2 A

3 B

4 D

5 C

6 78¢

7 35¢ more is needed to buy the bracelet.

8 He can buy 3 bags of crackers.

9 The football costs $6 more than the tiger.

10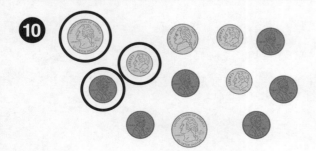

Answers may vary.

11 Emma has the most money.

12 Mei has 52¢ less than Emma.

13 Mei has 63¢ now.

14 The skateboard costs $80.

15 The drink cost 85¢.

Year-End Assessment

1 D

2 B

3 B

4 A

5 C

6 B

7 A

8 D

9 B

10 C

11 B

12 A

13 D

14 C

15 C

16

17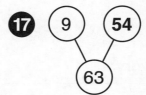

18

left   right

19 Mary buys 3 fewer <u>muffins</u> than <u>cookies</u>.

20 Mary buys <u>12</u> muffins, pretzels, and bagels in all.

21 She has <u>0</u> pretzels now.

22 58 + 9 = 10 + ┃ 57 ┃

Test A

Year-End Assessment

23 | 88 | 13 | 38 | (83) | 93 | 35 |

24
| A |
| **B** |
| C |

25
Answers may vary.

26
| 12 | 21 | 42 | 65 |
| 76 | 89 |

27
| 71 | **73** | 75 | 77 |
| **79** | 81 |

28 60¢

29 The time is 4 : 50 .

30 It is ten minutes to 5.

31 The teacher gave counters to 5 students.

32 91 – 3 = 88
There are 88 blue balloons.

33 33 (+) 25 = 58
58 pears are in the basket.

34 54 (–) 34 = 20
The chair costs $20.

35 He has 24 cookies.

Year-End Assessment

1 C

2 B

3 B

4 C

5 D

6 C

7 A

8 D

9 C

10 B

11 C

12 D

13 B

14 C

15 B

16

left   right

17

| 19 | 39 | 48 | 50 |

| 72 | 91 |

18 Dion needs to put <u>8</u> more green crayons into the box to have the same number of green crayons as red crayons.

19 There are <u>0</u> purple crayons in the box.

20 There are <u>30</u> red, blue, and yellow crayons in the box altogether.

Year-End Assessment

21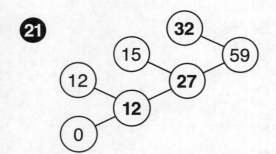

22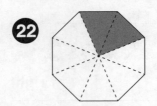

Answers may vary.

23 62 ⊖ 28 = 50 ⊖ 16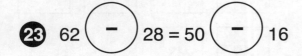

24 Pole A is <u>4</u> units shorter than Pole D.

25 <u>D</u> <u>B</u> <u>C</u> <u>A</u>

26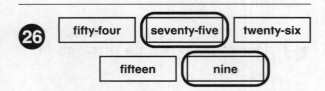
fifty-four | (seventy-five) | twenty-six
fifteen | (nine)

27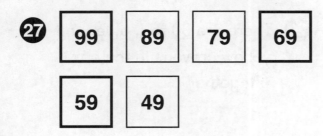
| 99 | 89 | 79 | 69 |

| 59 | 49 |

28 $26 + \boxed{39} = 73 - 8$

29 The pair of shoes cost $<u>42</u>.

30
There are 4 quarter hours in an hour.	
There are 3 half hours in an hour.	✓
There are 15 minutes in a quarter hour.	

31 <u>8</u> people are in line.

32 $\boxed{98} \;⊖\; \boxed{31} = \boxed{67}$

The farmer has <u>67</u> green apples.

33 $\boxed{51} \;⊖\; \boxed{13} = \boxed{38}$

Joe has $<u>38</u>.

34 Emma went to bed at

$\boxed{8} : \boxed{45}$.

35 He needs <u>4</u> plates.